THE *SAVVY* GUIDE
TO DIGITAL MUSIC

Other books in the *SAVVY* Guide series

The Savvy Guide to Home Theater, ISBN 0-7906-1303-4

The Savvy Guide to Digital Photography, ISBN 0-7906-1309-3

The Savvy Guide to Fantasy Sports, ISBN 0-7906-1313-1

The Savvy Guide to Home Security, ISBN 0-7906-1315-8

The Savvy Guide to Motorcycles, ISBN 0-7906-1316-6

THE *SAVVY* GUIDE
TO DIGITAL MUSIC

RICHARD MANSFIELD

PUBLISHING

Indianapolis

International Standard Book Number: 0-7906-1317-4

Chief Executive Officer:	Alan Symons
President:	Scott Weaver
Chief Operating Officer:	Richard White
Acquisitions Editor:	Brad Schepp
Editorial Assistant:	Dana Eaton
Copy Editor:	Cheryl Hoffman
Technical Editors:	Andy McEwen, Richard White
Typesetter:	Cheryl Hoffman
Cover Design:	Robin Roberts
Cover Photos:	Used under license. Mark Scott / Getty Images
Interior Illustrations:	Provided by the author
Interior Photos:	Provided by the author
Drawings:	Provided by the author

"Savvy Guide Icons" section adapted from *The Savvy Guide to Home Security* by John Paul Mueller.

Manufactured in the USA, at Ripon Community Printers, Ripon WI.

To my good friend Cliff Way,
who is both highly perceptive about music
and has an impressive musical memory

CONTENTS

Contents

SAVVY GUIDE ICONS

As part of Sams' Savvy Guide series, this book may use a number of icons to identify special kinds of information. These icons alert you to something unique, something that you should pay particular attention to as you read the book. Here are the four icons that may be used in this book and a description of what they mean.

 Savvy Tip: A Savvy Tip gives you special information—usually something that you won't find anywhere else. Often the Tip is based on the author's experiential knowledge, or as a result of an interview with an expert.

 For More Information: Sometimes you need just a little more information about a topic discussed in the book. Yes, the book described the topic well but possibly not in enough detail to meet your specific needs. This icon alerts you to additional information that you can find online. Normally, this icon points you to a detailed resource provided by a third-party resource.

 For Skeptics: Not everyone will accept the information in this book at face value; many people want proof. Although this book strives to provide as many details as possible, sometimes it will direct you to a resource that provides detailed proof about a concept. You can skip this information if you agree with what the text is saying; only the skeptic needs to read further.

Information identified by the magnifying glass icon is of three sorts:

- *Added Detail.* This is information that is nice to know but not essential for your understanding of the topic. In fact, the information might not even relate directly to the current discussion; it might have only a casual relationship to the topic as a whole. The ancillary details are sometimes important for a complete understanding of a topic.

- *Background Info.* Some people like to know the basis for a piece of information or the reason that the information is so important. Text identified by this icon provides that information, which you can use as a basis for additional research.

- *Elaboration.* Sometimes it's better not to include every detail in a discussion. This information elaborates on the current discussion, extending it in a way that completes the discussion for those who want to know more than the minimum required. You can skip this information if you're happy with the level of discussion provided in the standard text.

ACKNOWLEDGMENTS

I first want to thank acquisitions and project editor Brad Schepp for his thoughtful and helpful advice. He guided this project from the start, and in my view he's an editor who knows his stuff. I was also lucky to have copy editor Cheryl Hoffman asking the right questions and making so many good suggestions. And I also thank Dana Eaton for her positive attitude and her contributions to the completion of this book.

Technical editor Andy McEwen, head engineer at the Crucible recording studio in El Dorado Springs, Colorado, reviewed the entire manuscript for technical accuracy. His extensive experience as a professional recording studio engineer added considerable depth to the book, particularly in the more advanced chapters. I'd also like to thank my friend Chad Phillips for putting me in touch with Andy.

To these, and all the good people at Sams who contributed in so many ways to this book, my thanks for the time and care they took to ensure its quality every step along the way from original idea to final publication.

INTRODUCTION

Recently music fractured into millions of pieces. Or, as the professors like to say, music has been massively deconstructed. It's now in bits—meaning that computers (and you) can manipulate it in pretty much any fashion you wish. Whether you like to listen to and collect music, or want to modify or make your own music, this book has plenty of advice, tips, and techniques you're sure to find useful.

MANAGING YOUR MUSIC

The Savvy Guide to Digital Music is your handbook for all facets of the current digital audio scene. The book explains in detail—with plenty of hands-on examples—everything you can do with digital music.

We've now got incredible freedom to manage and control our favorite music in a dozen different ways. Here are a few topics covered in the book:

- Optimize and manage portable music players like the iPod.
- Capture and record *any* audio that you hear on your computer—streaming audio from Napster, satellite radio, news reports, and so on. Whatever you can hear, you can record in high-quality digital audio. There's even a free utility you can download that does this and more.
- Spread the music: broadcast music wirelessly from your computer—or portable player—with CD-quality sound. Several devices wirelessly send quality audio around your house.
- Deconstruct, then reconstruct, music using MIDI. Huge amounts of popular and classical music are available in MIDI format. With a MIDI-based sequencer you can manipulate syn-

thesized, sampled, or audio tracks infinitely. Recently Steven Tyler announced a new service that goes beyond MIDI. With the Umixit technology, artists will release their actual original tapes as separate tracks—vocal on one track, bass on another, string pad on another, and so on. Then purchasers can remix the song as much as they want, adding effects, substituting a their own vocal track, slowing down the chorus, whatever. This technology gives the average person the tools to remake favorite songs, tools that were previously only available to a few producers and engineers. For one thing, it perfects karaoke—you get to really sing while Eric Clapton backs you up on guitar. For another, you can use the original tracks to build mash-ups, remasters, and remixes to your heart's content.

- Add special effects such as reverb, chorus, EQ to your favorite songs. Or remove hiss and pops from your old tapes and vinyl records, then transfer them to CD for permanent storage.

- Discover the latest in digital audio, such as time shifting with podcasts, making mash-ups, accessorizing your portable music player, upgrading your sound card, choosing the best online music service, sampling, composing via computer, sequencing, and improving your listening room or recording studio the easy way.

WE'RE AT THE TIPPING POINT

Today's home computer is a music machine of enormous power and versatility—if you know how to use it. In fact, digital music is exploding. It's at a "tipping point," according to market researchers. Digital audio devices and software are about to flood consumers with astonishing new technologies. You'll pretty much have complete freedom when recording, storing, remixing, creating, and listening to music any way, any time.

Millions of people are buying iPods and other digital audio devices. Satellite radio is catching on, experiencing triple-digit sales increases. More and more people are realizing that they don't have to buy a whole CD to get one or two good songs—they can create their own CD's. A music studio with professional digital effects, multitrack mixing, and high-quality synthesizers and samplers can now be "built" inside a computer for a few hundred dollars. These same technologies would have cost far more only a few years ago.

These and dozens of other music applications are now possible for a small investment in money and time. This book demonstrates all the ways that computers can record, store, play, manipulate, and even compose brand-new music. No other book I've found covers this subject in its entirety. There are books specifically about the iPod, and others that focus on MP3 or setting up a home studio. But none explores all facets of today's digital audio.

EXPERIMENT FOR FREE

This book covers the waterfront, with practical, down-to-earth demonstrations and advice. Rather than waste your time with lengthy discussions about the theory of digital audio formats, I define them briefly, then illustrate the techniques and trade-offs with specific examples you can try on your own computer. For example, I show you how to hear the difference in audio quality between various sampling rates and file types, and then make informed decisions about the trade-off between file size and sampling quality. And I point you to lots of free music resources and demos, so you can download and experiment with some of the best music and music applications on the Internet.

With over 20 years of experience exploring digital audio, and the enthusiasm for consumer electronics of a rabid early adopter, I believe I'm well qualified to write a book that explains today's mushrooming digital music scene to the savvy reader. And this book's technical sections have been carefully edited by an experienced professional, Andy McEwen, head engineer at the Crucible recording studio in El Dorado Springs, Colorado. Andy's contributions are invaluable.

THE SKY'S THE LIMIT

This book tells you exactly how to distinguish between what's useful and what's merely a waste of time. I demonstrate that pretty much anything you want to accomplish musically is now possible for the average person. Whether you want to accumulate a mega collection of songs that rivals any radio station's record library, or aim to become the next Elton John—creating and recording brilliant original compositions—with today's affordable digital audio technologies, the sky's the limit.

Or, perhaps more accurately, your talent's the limit. There's more than one kind of musical ability. Maybe you can't sing in tune, but you've got golden ears and can create gorgeous mixdowns. Or you can come up with some great orchestrations (some Coldplay songs sound pretty cool when played by a string quartet). But you'll never know how talented you really are until you give it a shot.

And, until now, few of us had the opportunity to try some of music's most interesting jobs: conducting, adding effects, orchestrating, recording, arranging, mixing, remixing, sampling, mastering, and so on. This book shows you exactly how you can try your hand at all these aspects of music production—inexpensively, and with professional-quality results (if you're interested in spending some time learning the techniques, and, of course, if you've got the talent). But you'll never know until you try.

1 UNDERSTANDING DIGITAL AUDIO

◆ What digital audio can do for you

◆ Hear it now: streaming

◆ The dominant digital audio formats

◆ Working with MIDI

◆ Understanding compression

Digital audio has transformed music. Thanks to an explosion of inexpensive, yet astonishingly powerful, digital audio technologies, personalizing and manipulating music is now affordable and easy. With the possible exception of photography, music has benefited from digitization more than any other art form. If you can imagine it, you can do it.

However, most people don't yet realize that they now have virtually complete control over their music. It's not merely that you can collect and store thousands of songs on a device that fits in a shirt pocket. You can also get down *into* the music and customize the sound of existing songs—or create brand-new music.

For example, with a $49 program called PowerTracks it's easy to add a new drum track, or a violin accompaniment, or *any other instrument or music* to an existing song. You can add harmony to a melodic line, sing along with Norah Jones (then have the computer fix any off-key notes . . . yours, not Nora's), reorchestrate your favorite Bach fugue from organ to string quartet, or punch real gunshot sounds into just the right places in *Janie's Got a Gun*.

This book divides into two primary sections: the first part of the book explores ways to improve your listening, collecting, organizing, and otherwise managing your music. You discover your options among the many portable music players, explore ripping, burning, downloading, broadcasting, and other tips and techniques to enjoy music listening—whether at home or on the go.

The second major section gets you into music making itself, mixing or modifying existing songs, or creating new music of your own. Anyone with a good digital camera and a computer now has the tools to rival the best photographic developing and printing professionals of a few years ago. Likewise, today's amateur musician with a computer gets music-making "studio" time for free that would have cost thousands of dollars a short while back. You'll be amazed what you can do to Elton's songs, either by modifying them via digital effects, editing or creating a MIDI file, or recording your own new versions to your hard drive.

In this chapter you're introduced to the major technologies that underlie and drive today's exciting digital audio scene. After reading it, you'll understand the uses and techniques of streaming and the importance, strengths, and weaknesses of the primary digital music storage formats: MP3, WMA, WAV, and MIDI. And you're introduced to compression. Throughout the book, these topics are explored in more detail, but it's good to get an overview now of some of the main digital audio concepts.

STREAMING: AUDITIONING DIGITAL MUSIC

Internet radio stations and music stores allow you to audition tracks so you can decide whether to buy them. After you purchase music, you download the song to a file on your hard drive, but auditioning is more like listening to radio—the music isn't stored on your computer, it's *streamed* through your audio system.

A couple of years ago, it was often painful to listen to streaming audio. Before broadband Internet connections, the compression used in streaming lopped off highs, lows, and even midrange sounds. This often resulted in hollow, slowly pulsing, tinny, almost vocoder-like, low-fi music. But with today's high-speed transmission rates, streamed audio is quite listenable.

The advantages of streaming are that you don't have to wait for it to download to your hard drive, or commit to the music by paying for it, or use up space storing it.

Streaming works something like those shock-proofing antiskip techniques used in portable CD players. You click a link on a Web page to start the streaming process, and there's a slight delay as a bit bucket is filled with a few seconds of the streaming music. Then the sound starts pouring out. The bit bucket—a cache or buffer—provides insurance in case there's an interruption in transmission as the song comes over the Internet.

Nonetheless, you may have experienced pauses in streamed audio. This means that transmission slowed sufficiently to use up the buffer's contents and the audio was halted until the buffer could be refilled. Does this mean the buffer was too small? Yes, but it's designed to be the best trade-off between making you wait too long for the song to start and causing pauses when Web traffic or interference slows things down.

If you subscribe to an online music store, such as Napster (see chapter 8), the usual approach is to let you try out a song by listening to it streamed before you buy (download) it to store in your portable audio device or listen to at home via your computer. This auditioning can be rather like listening to a radio station: You have to follow the playlist, you can't simply search for a certain artist or track. Alternatively, if you already know what you want to buy, just use their search feature, typing in the artist or track.

However, not all services follow Napster's practice of allowing you to listen to the entire searched tracks as often as you want. Some don't let you listen to a searched track at all, others limit you to 30 seconds or low-fi. My suggestion is that you check out these options—and any monthly subscription charges (which are in addition to purchasing songs)—before signing up.

If you don't feel the need to permanently *own* songs, Napster, Real, and Yahoo offer unlimited download subscription services where you rent the music rather than permanently buy it. In other words, as long as you continue to pay your rental fee, you can download as many songs as you want (for listening only, not permanent storage). You synchronize your player with the online service once a month, or the songs evaporate. You *can* transfer songs to CD (thereby making them available for permanent storage and *ownership*), but Yahoo changes 79 cents per song, the others 99 cents.

Savvy Tip

Yahoo's online music service is undercutting the prices of competing services such as iTunes. Yahoo sells subscriptions for $60 a year or $6.99 a month, a new low price point. Yahoo's rivals, Napster and Real, charge $180 annually, or $14.95 a month. Many portable music storage devices can be used with Yahoo's service (Yahoo uses Microsoft's superior WMA format). Apple's iPod isn't compatible with the Yahoo service.

MP3: A STANDARD BY DEFAULT

Uncompressed digital music files, such as .wav files, are huge. They eat up hard-drive space in your computer or portable music player. Compression allows you to store far more songs on the same computer memory. And today's standard for music compression is MP3. Indeed, portable music devices are sometimes called *MP3 players*.

MP3 isn't the best way to compress musical data, though it's by far the most popular technology. It's not the most efficient. WMA, Windows Media Audio, produces files that are usually half the size of MP3 files. Nor does MP3 compression offer better-quality sound. Alternative compression technologies result in at least comparable, sometimes superior, audio.

For musicians and music companies, WMA is also preferable because it has a built-in DRM (digital rights management) scheme to attempt to prevent piracy. MP3, notoriously, does not—even though MP3 requires that people who sell software that incorporates MP3 pay a royalty.

The reason that MP3 dominates is, simply, that it dominates. It reached a critical mass several years ago, and now, if you want to sell a portable music player or a music-playback utility, or otherwise engage in digital song storage, you've got to at least include MP3 capability. Most people who have been collecting songs collected them in MP3 format.

Just as VHS overtook Beta as the format of choice for videotaping, so too has MP3 achieved dominance in contemporary audio storage—and that's not likely to change. Think of all the people who would have to use a converter utility to move their libraries of songs from MP3 to some superior format. And if the new format were sonically superior, they'd have to simply junk their current libraries and start downloading, and presumably paying, for a whole new song collection.

INTRODUCING MIDI

In this book you learn not only how to collect and listen to music but also how you can easily and inexpensively become actively involved in editing, recording, or creating it. To do this, you need to understand MIDI. It's a big part of today's digital music scene.

MP3 is a way of reducing the size of *audio* files. A one-to-one, uncompressed digital recording of a typical three-minute pop song takes up about 30 megabytes (MB). That's why you can only get a single album on a CD (which has 700 MB available storage, or about 70 minutes of uncompressed music). MP3 compression can reduce a 30 MB uncompressed song by a factor of almost 10, down to between 3 and 4 MB. MP3 storage—though compressed—still remains a *copy* of an audio file.

The iPod is the odd man out: by default it doesn't use MP3. Apple seems to prefer to lock its customers into its proprietary AAC format. However, if you're willing to spend some extra time, you can import CD music in MP3 format (which means you can also copy AAC songs to a CD, then copy them back in MP3). You can also translate WMA songs into AAC. For details, see the Apple iTunes Web site, www.apple.com/itunes/import.html.

Is there another way to digitally record music? Indeed there is—a very different way. It uses symbols to *describe* the music instead of digitizing a copy of the music's audio. It's called MIDI—musical instrument digital interface.

MIDI can reduce the size of a typical three-minute song to about 30 *kilobytes*—a reduction factor of 1,000. Of course there are some limitations: it only records instruments, not vocalists, who as yet have no electronic output plugs. Also, this format doesn't convert or compress an existing audio file. Instead, it records the music while it's played on a keyboard or other device designed to transmit the electronic signals that, collectively, produce a MIDI recording. MIDI doesn't accept audio from a microphone. It instead detects keys pressed on a MIDI-fied keyboard or other musical instrument.

MIDI provides complete control over electronic musical instruments (synthesizers, electronic drums, anything with a MIDI out jack). MIDI stores musical information in a highly compact format, and its data is easily edited.

UNDERSTANDING COMPRESSION

Why is a MIDI (.mid) file very small compared to the alternative, a recorded audio file, such as a .wav file? It only requires 6 bytes in MIDI to describe a note of *any* duration: 3 bytes for the Note On message, and 3 for the Note Off message.

If you press a key on a synthesizer and hold it for five minutes, you still record only those 6 bytes in a MIDI file, but it would require around 50 MB if you recorded that held note to an audio format, like a .wav file. Audio files record a waveform continuously; MIDI records only the details that change. MIDI rarely has to record a continuous, lengthy string of data (but it must do so in special situations, such as pitch bending or subtle changes in pressure on a synthesizer [synth, for short] key, for example, called *aftertouch*). MIDI is described in detail in chapter 10.

In one way, MIDI is similar to digital compression techniques. When you want to reduce the size of some data, the most obvious first step is to remove repetitive information. Think of a Warhol soup-can painting. If you divide it into perhaps 5,000 cells, you immediately notice that perhaps the first 189 cells are simply white. Rather than store "white" data 189 times, you can just specify a shorthand notation like "white x 189." Then, when you later re-create (decompress) the data, you have the computer display the original 189 white cells by translating the notation rather than merely copying a bunch of repeating cells.

A compressed video of a talk show, for example, often merely needs to record that nothing changed from the previous frame other than some motion involving the host's mouth. The entire frame is not recorded, just those areas that changed. This reduces the file size and—put another way—reduces the bandwidth (transmission requirements) of the video.

In a similar way, MIDI only records what changes—though indeed it sometimes must be highly specific in certain passages or with some kinds of music. MIDI does record nuances like vibrato, subtle variations in timing, velocity, and so on.

And MIDI doesn't store an *analogy*—an audio copy—of the sound waves but instead stores a mathematical description—a shorthand representation of the qualities that describe a song. MIDI is to an audio .wav file what sheet music is to a pianist in concert. MIDI describes the music, and the computer can read and reproduce the music that MIDI describes.

RECORDING AUDIO

Of course, you often need to record audio via a microphone. Singers don't have MIDI OUT jacks, and some instruments, such as saxophones and strings, are extremely expressive and are therefore difficult to reproduce via electronics. Typically, today's sequencers—software that records and plays back musical events using MIDI—now allow you to store both MIDI and waveform (audio) tracks in the same project and mix them for a final audio mixdown (see figures 1-1 and 1-2).

The problem with recording sax and some other acoustic instruments via MIDI doesn't derive from any inherent limitations in MIDI itself. Indeed, MIDI has provisions for describing many kinds of musical behavior (you can play samples of breath, guitar squeaks, and other artifacts that make synth music or samples sound more authentic). The problem is that it's difficult to "play" a sax on a MIDI keyboard. Keyboards don't have accessories to imitate various sax players' techniques, nor is it easy to build a MIDI sax that could capture all those subtle sounds and translate them into symbols.

But a MIDI file is nonetheless quite flexible and powerful. It can contain many instrument tracks simultaneously (each instrument's data on its own channel), and you can easily edit each little "event" and reproduce the original performance precisely.

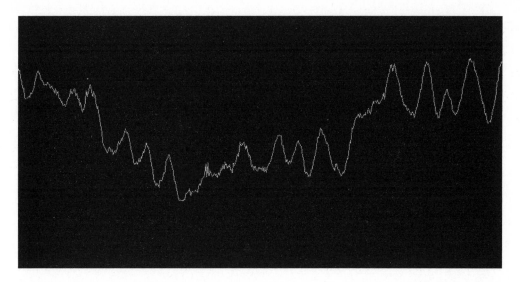

Fig. 1-1: An analog audio recording is often stored in a .wav (wave) file format, and the format is well named. As you can see in this analog track, analog audio is a series of waves.

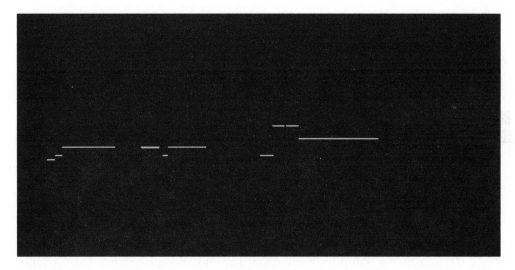

Fig. 1-2: Here's a MIDI track, and it looks digital—discrete quanta of data—particles, not waves, as Einstein would have said.

Now that you've got a feel for the major digital audio technologies, it's time to consider what's coming down the pike. The first question you might want to answer is: Should you buy some digital audio product you're interested in, or wait a few months until something newer comes out? After all, digital audio is currently one of the hottest, most rapidly changing technologies. What's new today is upgraded tomorrow. Should you, can you, wait? Chapter 2 considers the nature and pace of audio technology, and at the end advises you to go ahead now and buy some great digital audio device.

2 FUTURE TECHNOLOGY

◆ The explosion continues

◆ Digitization equals freedom of choice

◆ Reaching sufficiency

◆ Quality remains subjective

◆ Wirelessness spreads

◆ Ultra wideband: the ultimate solution

◆ Should you wait a few months to buy new equipment?

Apple has trouble keeping up with the demand for iPods, and it seems that each week you can read announcements about new recording software, improved portable music devices, more sophisticated surround sound mixers, and all kinds of other digital audio gear. Personal computers burst onto the consumer electronics scene in the 1980s, selling in huge numbers, and also improving rapidly in quality as they fell in price. A similar high-tech explosion is now happening to digital audio.

Satellite radio, music downloads, podcasting, affordable personal recording studios, and many other rapidly growing facets of digital audio can be summed up in one phrase: audio how you want it, when and where you want it.

In other words, a vast increase in cheap digital storage media, coupled with ever more powerful audio technologies, is resulting in a great increase in freedom of choice for music lovers.

That freedom expresses itself in many ways:

- Instead of waiting for the DJ on the radio to play a particular song, you've got the radio station's entire library of recordings right there in your iPod.
- Rather than wishing Bob Dylan played "Mr. Tambourine Man" just a little bit slower, you can just click a tempo button in a sequencer and play the song as quickly or as slowly as you want (without distorting the pitch and getting the "helium voice" effect).
- Instead of wondering what it would be like to be onstage, right in the middle of a string quartet rather than out in the audience, you can buy a high-definition surround recording that puts you smack between the cello and violin.
- Create a mash-up, the latest rage in remixing. Mix the vocals from one track with music from a different track, often from a song of an entirely different genre. The results are, well, a matter of opinion. Typically, a mash-up attempts to at least match tempos, lengths, and pitches— all of which the computer can do.

THE DEMOCRATIZATION OF MUSIC

Democratization is another way of describing the tremendous freedom that digital audio offers. Anyone can promote and distribute their own CDs online. Once there were only a few major record companies; now there are tens of thousands of labels.

Even music criticism is becoming possible for anyone with access to an Internet site or interest in creating a blog or podcast (see chapter 5).

Recording, mixing, DJ'ing (all technically at a professional level of quality), creating custom playlists, burning personalized CDs for friends—everything about music seems to have become possible for the average person. Talent, of course, is not evenly distributed, but that's about the only element of music creation that you can't buy today.

Niche and independent music is now much more popular and available. The audiences for these bands are small but can be intensely loyal. So the trend is similar to what's happening with cable TV: rather than having the traditional big-three networks determine content, there are now many more channels, each more narrowly focused and presumably more satisfying to their audience because they are more tailored to that audience's moods and tastes. With music, there are now many channels of distribution and promotion. *Everything's* available, and can be auditioned immediately, online. Also, there's no space problem online as there is in a traditional music store. So you can check out millions of tracks, not just hundreds of albums.

Here's another example of how digitization offers freedom of choice. Album sales have all but dominated music sales for decades. That era is ending. Now with iTunes and other music stores selling individual tracks, we're back to singles again (and this time there's no B side that comes along for the ride).

How did we get to these new ways of making and listening to music? The desire to have more control over our music has been there since the first drummer hit a rock with a stick. But only recently have technologies advanced sufficiently to blow music into billions of bits and give us all access to them. It's a matter of technological sufficiency.

THE POINT OF SUFFICIENCY

Technology moves forward relentlessly, and with increasing momentum. But for each type of content there is a *point of sufficiency* that, once crossed, results in no further significant improvement. For example, several years ago that point was passed for word processing. The first word processors suffered from a delay: if you were a good typist, you could get ahead of the screen, typing words that were not yet visible on the monitor. Today, though, no matter how fast a typist you are, the computer can keep up, storing the characters and displaying them with no problem.

A similar point has not yet been passed for recording music, but we're close. You still *can* play a sequencer keyboard faster than some sound cards can process the music; there's a delay between key presses and hearing the notes. This is called *latency* and it's a highly annoying effect. Fortunately, soon computers will be powerful enough to pass that point of sufficiency and the latency effect will be a mere memory.

It Can Take Decades

We have passed several points of sufficiency for music *listening*, but it has taken decades. Consider the experience of listening to a record. Perhaps you want to be able to sit down and listen uninterrupted to an hour-long symphony. This wasn't possible until recently. The symphony was always interrupted because the recording medium simply didn't have sufficient storage capacity.

Before 1950, a single symphony required a whole stack of 78s (they spun at 78 rpm). Every few minutes the music stopped and you had to turn the record over, or put on the next one. Some record players had a mechanism that permitted you to stack the records, and each would fall onto the platter as needed. It was claimed that this didn't really *hurt* the records (much) because the "air

ULTRA WIDEBAND: THE FUTURE OF WIRELESS

As you've surely heard, nasty hissy fights break out now and then over TV and radio "air." Anybody who gets control of a TV channel can make tons of money selling ads. But the heyday of local radio and TV stations has passed.

We've become used to having hundreds of TV channels. And we're getting used to that kind of freedom of choice for audio as well, thanks to several technologies such as podcasting, portable audio, and satellite radio. What's more, people want two-way freedom of communication, such as on-demand movies and, pretty soon, on-demand, TiVo-style radio and music.

These freedoms require sufficient bandwidth—the airspace and airtime—to transmit digital audio to hundreds of millions of people. Wi-Fi (the current 802.11g format) manages to transmit at 54 megabits per second (Mbps), but in most homes and businesses—thanks to those inconvenient signal blockers such as walls and floors—you get less than half that throughput, an average of around 23 Mbps.

What we need is something faster and better. Good news! It already exists. It's called ultra-wideband (UWB), and it manages a blinding 11,000Mbps.

UWB is fascinating; it's very different from familiar radio and Wi-Fi transmission techniques. UWB employs high-speed impulses rather than traditional radio waves. It was first discovered in 1962, but research on it was immediately restricted by the government. UWB didn't emerge back into daylight until 1994. Since then, R & D has been vigorous. UWB equipment is now, at long last, coming to market in 2005. See Intel's description of the technology at www.intel.com/technology/comms/uwb.

Bear with me on this; you'll be happy to be the first on your block able to explain it. A UWB device sends sub-nanosecond pulses. These are quanta of information, rather than classic wave-based transmission (like TV, radio, and pretty much everything else that goes out "over the air"). UWB pulses are very brief and require hardly any power (a few milliwatts, about 1/10,000 the power required by a cell phone). Information rides on UWB pulses, which are time modulated: A pulse that arrives at the antenna just a little earlier than it should represents digital 1. Digital 0 is a pulse that's a tiny bit tardy.

Although capable of long-distance transmissions at 11,000Mbps, the first commercial devices approved by the Federal Communications Commission are likely to be much slower and shorter than their potential. For one thing, the initial applications will not need warp speed or long distance.

UWB offers some additional interesting capabilities. It doesn't interfere with other signals (it's not like them, and so is not competing with them). It doesn't even require any bandwidth to be allocated to it—that precious airspace. It's ultrawide bandwidth, meaning that it scatters its data across the entire frequency spectrum rather than requiring its own, particular band within that spectrum like traditional radio, TV, and other transmissions.

What does this mean? The airwaves have been subdivided by the government into thousands of "bands": a band here for satellite TV, a band there for truckers' CB, yet another band up there for FM radio, and so on. It's as if you had a highway with 30,000 lanes and the FCC decided where to lay down all the dividers to prevent bedlam and interference. But believe it or not, we're running out of lanes! Hence the hissy fights.

UWB has nothing to do with these lanes. It's no more invasive of ordinary traditional media transmissions than a gentle breeze is of automobile traffic. Cars and soft winds are simply so unalike in size and speed that they might as well be in different universes. Even though technically they do "share the road" physically, they never get in each other's way.

UWB emits very high-speed "excitations," spurts of data so fast and alien that they fly right through the other, slower, traditional communications without affecting them at all. Like neutrinos through a mountain, UWB emissions are ghostly.

cushioned the fall." No wonder these records were quickly filled with crackles and pops. Worse, each time you played them, a needle dug into them.

Then the revolution: The LP (long play) record appeared just after World War II. It was a wonder because a symphony could fit on a single disk—but you still had to get up and turn it over halfway through.

Symphonies Fit Perfectly

Symphonic recording crossed its line of sufficiency when the CD arrived. With a playing time of 74 minutes per side on a CD, you could now listen to the entire composition without interruption. The whole thing fit on one side of its medium. Any further increase in storage size is merely a matter of multiplying the number of symphonies you can store.

I'd say that portable song storage has reached its sufficiency. When you've got, say, 10,000 songs in your iPod, that's enough freedom of choice for a few weeks, anyway. Then you can reload it.

Length, however, is a quantity. What about quality? I doubt perfect sufficiency will ever be possible by that measurement. It's said that when the first disk record was introduced (replacing wax cylinders), they set up Edison's 1877 record player behind a curtain. People *swore* they couldn't distinguish that music from real musicians. Later on, as people became more accustomed to the campfire crackle you can hear on any pre-CD disk, the bar for realism in recording was raised. Indeed, when CDs were first introduced, people frequently claimed they sounded just like live music. And HDTV reviewers often say that high-definition TV is so real that it's like looking through a window.

But as CDs demonstrate, reproduced music can be freed from noise, flutter, distortion, crackles, pops, and other glaring defects, yet the final mixdown will always be a matter of personal taste. Should the singer's voice be more prominent, or less, in the mix? Should the drums have been recorded with more mics? It's an endless debate. And even if you go to a live concert, somebody will cough or answer a cell phone. No sonic experience, recorded or real, can please everyone. With quality, sufficiency never seems to arrive.

For example, surround sound speakers seem to many people near qualitative sufficiency. But I disagree. What if a movie director wants you to hear a bomb whistling down on you from *above*? Even if you have 7.1 speakers (seven regular speakers, plus a subwoofer), they're still pretty much on the same plane—usually arrayed in a band around you at about the level of your ears. They've not yet gone spherical, adding more speakers above and below the listener until you're entirely within a 3-D bubble of controlled sound.

And even if you sit within a sphere of sound, you can always wish for better-quality speakers. Picky, picky.

Why Now?

If digital audio is exploding, why isn't digital *video* taking off as well? In terms of passive entertainment—watching video—the DVD is pushing analog videotape right into extinction. But in terms of active involvement (personal video creation), the technology isn't quite there yet. Personal hands-on video will take off, but music has simply been in the digital domain longer than video. The reason is that audio requires less power and space in computers, so it's easier to manipulate audio at this stage in computer evolution. Video makes greater demands on CPU processing power and storage.

The year 2005/2006 does seem to be the year of digital audio. Memory, CD-writer, and hard-drive prices have fallen enough; music software is powerful, mature, and sophisticated enough; the Internet offers easy, immediate, and cheap promotional and distribution channels. These and other reasons converge to create a perfect wave for digital audio to ride.

Video will eventually reach its point of sufficiency—it's moving toward total digitization—but it's simply not as easy to work with yet as audio. Commercial music recording studios are dying off—they're becoming irrelevant. Once upon a time, only movie stars had home theaters. Likewise, a few years ago only rock stars had home recording studios. Now every garage band can buy a high-quality personal studio for less than $1,000. And you don't even need the studio—most of what you need can be installed on a laptop computer.

Even some professional producers and engineers are simply setting up computer-based home studios rather than bother with the professional studio hardware they thought essential a few years ago.

Video production studios are still busy, but they, too, are threatened with future obsolescence once hard drives and RAM memory are large and cheap enough and computer CPUs are fast enough to achieve a point of sufficiency that renders inexpensive consumer equipment the equivalent of commercial equipment.

UP INTO THE AIR

The ultimate in audio sufficiency will be very small, highly convergent, and wireless—a single device that

- Receives and broadcasts all media
- You can easily carry
- Gets any content, at any time, anywhere you roam. Feel like hearing Alison Krause? You've got it instantly.

There's no more need to download, store, search, or otherwise manage entertainment like songs or symphonies. They're in the air, always and everywhere.

We're obviously not there yet, but Philadelphia is going Wi-Fi citywide. That's a start. And hundreds of channels of digital radio are beaming down across the entire country 24/7 via Sirius and XM.

We'll never have enough bandwidth to achieve total on-demand wireless two-way audio, you say? Good point, but wrong. Wi-Fi is only the start (see the sidebar on ultra wideband on page 28).

SHOULD YOU WAIT?

No. If you're looking to get a bargain—more power for less money—today's digital audio *already is* a great bargain. Wait a few months and you'll doubtless see even better bargains, or more attractive hardware, but so what? Do you want to sit around for months doing without? Like a jerk?

Sure, maybe the portable player you're thinking about lacks the ice-cool iPod style. For some reason Apple has always understood that many consumers want stuff that's not only functional but also *looks stylish*. Target knows this. Sony knows it (but sometimes forgets), though their new Network Walkman is sleek enough. And other manufacturers are slowly catching on. Maybe if you wait a few months, your favorite will look less froggy and more princely. If that happens, you can always upgrade. Sell the old one on eBay, right? It's the greater-jerk principle in action. Pass along your castoffs.

As you'll see throughout this book, the digital audio available today is very impressive across the entire spectrum of music technology. So my advice is: go on a shopping spree right away.

3 GOING PORTABLE: iPOD AND THE OTHER PLAYERS

- ◆ Choosing the right media player for your needs
- ◆ Where will you use it?
- ◆ Chips or disc
- ◆ Player audio quality compared
- ◆ Sampling rate and bit depth
- ◆ Considering compression
- ◆ The great battery controversy
- ◆ Nonmusical uses: audible books, data, photo, and videos

They're everywhere: iPods, MuVos, Zens, Lyras, Nitruses, iRivers, Chibas, Calis—you name it, they're selling fast. But what portable storage device is best for you?

Do you really need 40 gigabytes' worth of songs to carry around with you? Perhaps. It's sure convenient to have your entire music collection sitting there in a plastic cube the size of a pack of cards, ready anytime you want to listen.

This chapter is for those who are ready to upgrade or who've not even yet taken the plunge and bought one of these beauties. You'll see how to separate the hype from the truly valuable features, so you can get the most bang for your buck.

Savvy Tip

Always beware the hype. Ads sometimes stretch, if not actually rupture, the truth. For example, there are many complaints that a famous media player claims to handle photos but in practice is a less-than-satisfying photo management system (it's more a photo *display* system). The best advice: before you buy, type "*name of device* review" into Google and read some online wisdom in the electronics magazines. Also take a look at epinions.com and search there for your prospective device to see what average owners feel about their unit. You get some pretty honest, and usually helpful, reports from the field. After all, they've had it for a while and figured out what they like and what they wish were different.

First, we can't call them *music players* because some of them can store and display other kinds of information in addition to music, from computer files to video. But what they have in common, and what distinguishes them from simple portable file-storage memory devices, is that they are designed around software that manages media. Some can record from internal FM radios or record your voice via a small microphone. Others can play back recorded books from Audible, or shuffle the order that songs are played, or handle lots of compression formats such as WMA and MP3. One can even download Outlook data so you can view your calendar and tasks (Creative's Zen Micro). But they all do media, so let's call them *media players*.

This book is about audio, so if you're interested in getting a media player that also manipulates photos or something, take a look at the specs advertised by the manufacturer. Manufacturers are not shy about telling you all the great things their little wonder machines can do.

Make a list of the things you'll be doing with your portable player. Jogging, storing huge graphics files, plugging it into a quality stereo system, taking it into the jungle for two weeks at a time, whatever. For jogging you want something that is sturdy, for bulky files you need lots of memory, and for serious listening in a quiet environment you need the best possible compression scheme (and also look for reviews that stress the great audio quality of the unit). Jungle trips require long battery life or easily replaceable batteries.

THE TWO CAMPS: MECHANICAL VERSUS SOLID STATE STORAGE

Portable media devices store data in two primary ways: on a spinning mini–hard drive and in solid state (no moving parts) memory, like computer memory chips. The former devices hold lots more data but eat batteries and have mechanical parts that can fail. The latter are wonderfully light and

Four hours doesn't sound like a lot of music to those who want to store enough music to fill a radio station's library, but for many commuters four hours is more than sufficient. Sucking new music into the machine takes only a few seconds of your attention, so it's not even as tough as making the morning coffee. Seriously, don't just jump at the biggest storage out there if that's not what you really need. And do read about the battery issues in the section later in this chapter.

can run for 20 hours on a single AAA battery, but if you want entertainment that lasts as long as the battery, you're going to be limited to talking books, not music. That's because 20 hours of music requires much more storage space than a solid state device allows.

An example of the no-moving-parts camp, the 128 MB Creative Nomad MuVo can hold four hours of WMA music, two hours of MP3, or about 16 hours of good-quality recorded voice. One little MuVo gem goes for under $50 at Wal-Mart and is beautifully designed. The size of a disposable lighter, it pulls apart, revealing that one end of it is a USB plug that you just stick into the nearest computer and drain out some Mu or Vo for later listening.

For a little more cash (about $150 street) you can switch camps from solid state (chip) memory up to rotating disc storage—and get a 1.5 GB hard disc–based Rio Nitrus that can hold 50 hours of WMA music. Its rechargeable battery claims 16 hours continuous play, and for good measure they throw in a five-band equalizer so you can fiddle with the contour of the sound.

However, the really gigantic, megastorage units cost quite a bit more, the most famous being the iPod 40 GB unit that can hold a lifetime library of 10,000 songs ($399). It claims "up to" 12-hour battery life, which may be like my claim of lifting "up to" 300 pounds. That was on Valentine's Day in 1978, and I've never fully recovered. I was moving a stove for my prospective mother-in-law while everyone watched.

Let's not call these special cases or anything, but some battery-life claims are, how shall we put it? Optimistic. Like they would manage to achieve that battery life in a hyperbaric chamber on the space station. Here on Earth, batteries sometimes don't tend to last as long as the ads claim, though I've not tested the iPod, so perhaps their specs are even conservative.

CONSIDERING AUDIO QUALITY

Analog recording looks like, or *is analogous to*, the music it records. If you could make yourself very small and walk along a groove in a vinyl LP, you'd see small rippling waves in the groove walls for a cymbal hit and a few, great, thick, long waves for a bass-drum thump. But whatever size the grooves are, the waves along the walls would be continuous.

By contrast, digital recording looks nothing like the original music. Walk along a CD surface and you'll think you're in a Star Trek alternative universe. You see an infinite swirl of pits across a flat rainbowed mirror. Those pits give it away: this stored sound is not continuously imitated like the waves on a vinyl record but instead is sampled at some specified interval. Samples are quanta. One important side effect of quantization is that it's always necessary to *infer* sound in the space between the pits. Some of the sound is missing, no matter how often you sample the original wave.

Consider how visuals look when they've been sampled. A display of quantized (digital) information is not the same as the analog reality you see around yourself every day. If you walk up real close to most TV screens, you can see the little, separate, mosaic-like units (pixels) that make up the grid pattern of the picture. But if you walk up to a tree or flower in the real world, no matter how close you get, you don't see any grid pattern.

Even if technology manages to reduce the digital grid beyond human perception, digital displays will nonetheless always be quanta. Sometimes the grid gives itself away spectacularly. That's why if you go on TV they tell you, please, don't wear a tie with diagonal lines in it! A grid pattern is horizontal and vertical by nature, so when you violate that with diagonal information, artifacts such as moiré patterns, rainbow false color effects, and dot crawl are the result.

It's obvious that because you *are* taking samples, you cannot get a *perfect* copy of the original. There must be gaps, no matter how many samples you take. But on the other side of the coin, analog recording techniques such as tape recording suffer from their own set of problems, mainly related to noise. Analog techniques cannot copy exactly what they hear either because there's always at least some noise introduced by the medium itself (hiss on tape, for example, or that crackling and popping sound on old vinyl records). There's also mechanical variability in the transport mechanism (causing effects such as flutter, and other false variations in pitch and volume like wow, which is slower than flutter, but annoying nonetheless). Also, analog storage ages badly, eventually breaking down entirely, such as when whole flakes come off a VHS tape.

It's impossible, so far as we know, to make a perfect copy of music.

Nonetheless, the more information (the more often you sample the original sound), the more accurate a digital reproduction will be later. But more frequent sampling also means you have to pay the price of needing more storage, or, put another way, greater bandwidth when you send the music from here to there.

Digital music is usually *compressed* so you can pack more music into the same memory, or flow more through a transmission pipeline like an Internet radio station or your local wireless network. However, many attempts to compress cause at least some loss of quality. Before considering compression systems in more detail, first consider why compression is necessary: digital audio is stored as numbers, numbers represented by pits on a CD surface, or by bits in computer memory, or by magnetic patterns on a hard drive. Whatever the storage medium, data takes up room. How much room depends on the quality you want to achieve: how generous is your sampling rate, and how large is each sample (the *bit depth*, or how many bits are used to store a single sample)?

UNDERSTANDING SAMPLING RATE AND BIT DEPTH

When you record something digitally or compare the audio quality of various digital audio playback formats, there are two primary specifications that distinguish the various formats: *sampling rate* and the size of each sample, or *bit depth.* The sampling rate tells you how often per second (hertz, or Hz) the sound is sampled: 44,100 Hz is CD-quality, but newer formats such as SACD (Super Audio CD) and DVD-Audio sample at twice that rate. Hz is generally divided by 1,000 and expressed as kilohertz (kHz) in specification sheets; Thus, 44.1 kHz for CD. Although 44,100 Hz can record the frequency range that the best human ear can detect, as anyone who has programmed computers knows, digital numeric storage is subject to rounding errors, and extra "precision," as it's called, is made available when you give yourself larger numeric ranges to store your data. (By contrast, AM radio tops out at around 8,000 Hz and FM at 22,050 Hz.)

The second primary digital audio specification is the size of each individual sample, and here, again, more is better. Called *bit depth* (the number of computer memory bits used per sample), this measurement is 16 bits for CD, allowing a number ranging from 1 to 65,537 to be stored for each sample. SACD and DVD-Audio generally employ 24 bits, or a range of nearly 17 million. Any kind of sonic manipulation—remixing, adding reverb, whatever—adds errors into the digital domain. Having a greater bit depth reduces the impact of these errors when the signal is eventually restored to analog form and pumped as waves out speakers to our ears. Converting the waves from a microphone's vibrating diaphragm into a stream of numbers (sampling) is called analog-to-digital conversion (ADC); going the other way, translating the numbers into waves for speaker output is called digital-to-analog conversion (DAC).

Now you see what we're up against. High-quality SACD music storage requires 3 bytes of storage (24 bits) multiplied by 96,000 *to store only one second of recorded music.* Every four seconds of SACD music requires over 1 million bytes of storage.

CD quality is half that, but a minute of CD music still needs over 10 megabytes. That's why a CD-

R allows you about seventy-five minutes of CD-quality music (a CD can hold a bit less than 750 MB). Put another way: one CD = one Sheryl Crow album, with maybe a "bonus" track thrown in.

Obviously, this size limitation is merely temporary, just one phase in the continually collapsing cost of digital storage. Eventually you'll be able to squirrel away every movie and song ever recorded on a wafer the size of a Ritz cracker. At some point in the future we will no longer need to compress music. But until the next leap forward, when blank 50 GB double-sided Blu-ray blank discs are selling for a penny each in Circuit City, people want to carry around more music than the current CD format is designed to store. Enter compression.

DIGITAL COMPRESSION SCHEMES

Here's how the simplest compression works. If for a second or two all music stops in a Maroon 5 song (as it sometimes does—Shania Twain is very fond of this sudden silence technique), why record the silence? Two seconds of recorded nothing at CD-quality can take up about 125,000 bytes! How about reducing it to a few bytes that say, in effect, "Make no sound for two seconds"? Often you can *describe* repetitive information—its nature and duration—far more economically than you can record it. Audio and video recordings always include at least some repetitive information, so there's no reason to record all those repetitions. Just go through and analyze the original, uncompressed recording to see where you can insert shorthand descriptions in place of recorded data.

The most popular compression format by far is MP3. It has been around for a couple of decades but recently became the de facto standard. It's not the most efficient compression, nor does it produce the best-quality audio, but given its dominance and wide usage, it's likely to remain the technology that most people use and, more importantly, that most digital media employ.

MP3 files are usually encoded using a constant bit rate (CBR) because it's the simplest approach and most people don't realize there's even an alternative. But consider real-world music: It's rarely consistently soft or consistently simple (with few instruments and few notes). It gets loud or complicated at times, though it can be simple at other times. The bit rate that has more than enough headroom to handle simple, quiet passages has to strain to compress and decompress loud, complicated music. This results in perfectly listenable passages in some of the music (such as the gentle start of an Enya minisymphony) but causes nasty distortion when she really gets going later on in the song. A good solution would be to vary the bit rate, which is exactly what VBR (variable bit rate) compression achieves.

Of the alternatives, the famous LAME MP3 VBR encoder works reliably and well. It listens to the music and adjusts the bit rate from 32Kbps up to 320Kbps, though this range can be modified by

There's an unfortunate disconnect between the measurement of frequency, or digital sampling rate (expressed in kHz, or 1,000 cycles per second), and the measurement of bits per second (Kbps, or 1,024 bits per second) typically used to describe digital *transmission* rates. When describing compression bit rates, you'll normally see the rate expressed in Kbps, with 190Kbps a common figure.

the person recording the music. Continually upgraded and widely supported by open source audiophiles, LAME currently offers recording default settings ranging from an average of 190Kbps up to the MP3 maximum quality of 320Kbps. The usual trade-off is between higher quality (resulting in big, fat disk files) or lower-quality (more-compressed, smaller files). However, you can also trade off speed of encoding sometimes for a loss in quality as well.

GET RIPPING

Let's take a quick look at one popular ripper, Windows Media Player 10, to get a feel for your options when ripping track off a CD. In Windows XP, choose Start|All Programs|Accessories |Entertainment|Windows Media Player. Violating their own long-standing rules about predictable menu location across applications, somebody at Microsoft decided it would be cute to really hide the menus in Windows Media Player 10.

Finding the Menus

But we *want* to get to the options on the Tools menu. Here's the secret: Locate the small down-arrow symbol in the top right corner (next to the underline symbol that minimizes the window). Click that down arrow and there they are, the cleverly hidden menus. Choose Tools|Options and click the Rip Music tab in the Options dialog box. You see the dialog box shown in figure 3-1.

By default, Microsoft's WMA compression technology is used, but you can select MP3 if you prefer. WMA achieves greater compression at comparable quality to MP3, and WMA is widely supported by many digital storage devices as well. Devices typically list WMA storage capacity as double that of MP3. For example, the SanDisk SDMX1 256MB unit is said to hold over four hours of MP3, but eight hours WMA.

WMA is also the darling of music companies, and of many musicians, because it offers a technology—digital rights management (DRM)—that can prevent people from making multiple additional

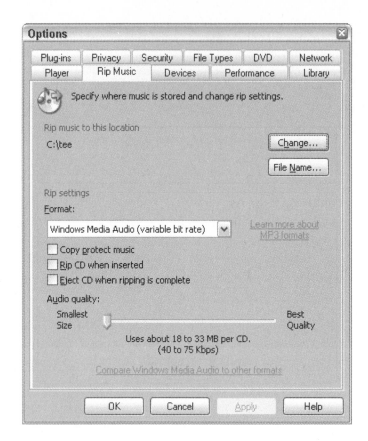

Fig. 3-1: Here's where you can adjust the rip settings in Windows Media Player.

copies of a song. Apple's iPod uses MPEG-4 AAC (advanced audio coding), and it too includes a DRM scheme.

Three Versions of WMA

Notice if you drop the listbox under Rip Settings, Format in the Media Player, you're offered three versions of WMA: standard (CBR), variable (VBR, the default), and lossless. If you choose the last of these, you're not allowed to adjust the Audio Quality slider: it simply slams itself up against the Best Quality pole and disables further movement. The lossless setting is a VBR technique that ranges from 470 to 940Kbps. The other WMA formats are adjustable, and as you move the slider up toward the Best Quality setting, you can see the file sizes grow. A typical CD ripped via WMA VBR at the best-quality setting results in a file between 100 and 150 MB. So when you go to save this to a blank CD, you can store between five to seven CDs' worth of compressed music. The lossless

WMA setting cuts that storage in about half. The MP3 option is not variable bit rate and results in around 50 to 150 MB per recorded CD, depending on how you set the Quality slider.

To rip, just select the format and set the quality slider, then put in a CD. Click the Rip Music button in the main Media Player window to start the ripping. The resulting files are saved to your default temporary directory, though you can change this location in the dialog box shown in figure 3-1.

Compression Format

What storage method should you use? As usual, it depends on your circumstances. My best advice is to always record original audio—such as little Suzie's recital—at the highest possible quality (namely .wav files or on DAT tape or some other uncompressed format). You can always afford to store these original files on CD or DVD, with blank CDs on sale costing less than a penny each. But you can never turn back time and improve on a cheesy, lossy original recording.

As for transportable music, if you're listening via an inexpensive pair of earphones, on a noisy subway, or while jogging, or if you're using a device with little memory, you might well find the WMA constant bit rate at a 128Kbps setting, with its small file sizes, provides perfectly acceptable audio quality. Or if you're old—say beyond thirty—and your hearing has deteriorated. Or if you're young, but have foolishly whacked your ears with too much volume for too long . . . well, you get the idea.

On the other hand, if you have great ears or one of those giant portable hard drive units, or if you plan to store your music on a home server for play through hi-fi via Wi-Fi, you'll perhaps want to go for a lossless format.

But do factor into your thinking the history of technology: storage media rapidly drop in price and expand in capacity. And although surround sound formats will require additional storage, do select the best possible format, the best possible quality, when *archiving* your favorite music or kids' recitals and the like. How you then compress it for portable listening is another matter and is governed by your intended uses. But always archive and record originals using the highest quality possible.

Battery Problems

One major complaint heard about the iPod is its rechargeable battery's life. This isn't the duration between charges but rather how long you can use the battery before it needs to be replaced entirely. You can read about allegations that Apple's battery stops functioning after about eighteen months and must then be replaced at www.ipodsdirtysecret.com.

Apple now offers a $59 out-of-warranty battery replacement, described at www.apple.com/support/ipod/service/battery.html (or call 1-800-APL-CARE), and a $59 extended warranty program.

Savvy Tip — Your circumstances dictate what kind of battery life you need. You can use one of the little AAA-battery-driven units for maybe two or three straight days nonstop, and keep a couple of replacement batteries in your pocket. Or, if you're near an electrical outlet often enough, it doesn't matter if your hard drive unit runs out of steam after a few hours—just recharge it periodically as required.

Apple certainly isn't alone in the battery complaint department, but for many customers of the various devices, it's often a question of the believability of battery-duration claims. Many people find that *in practice* (also known as reality), batteries that are claimed to last 20 hours between replacement or recharging actually often don't last that long.

Do remember to read online reviews at epinions.com or in online magazines (see the next section, "Making the Decision") if battery life matters to you (either the life between charges, or the time you've got before a rechargeable fails completely and needs to be replaced). Some manufacturers' battery-life specs are perhaps more hopeful than realistic.

MAKING THE DECISION

Books have longer shelf life than magazines, and magazines last longer than information on online sites. That means that for specific advice about the latest digital audio device models, you should go online and read reviews (after you've absorbed all the wisdom in this chapter):

- One good place for info on portable media is http://mp3.about.com/cs/portableplayers/tp/portables.htm.
- CNET also provides lots of good reviews: http://reviews.cnet.com
- And the old standby for man/woman–in-the-street opinions is www.epinions.com.

Audible Books, Data, Video

A bit is a bit is a bit. It doesn't matter to storage devices what kind of digital information you send into them for safekeeping. Movies, Bach, Beautiful Garbage, recorded books, your efforts at writing a novel—they're all the same to a bit bucket. What differentiates media players from each other is the software that interacts with you when you record, manipulate, or replay that stored data.

Some portable devices are capable of handling the security and user interface requirements of

Audible, a service that provides recorded books in your choice of compression rates. Audible books are very well produced and well read and include some of the most interesting current titles.

It's my view that their software—particularly the download and USB transfer aspects—can look forward to improvements. But because Audible allowed me to store and listen to a big, 12-hour-long, 456-page biography of Howard Hughes, with quite good quality playback, in a little Bic-lighter-sized MuVo, I love it.

It has solved the messy problem of having to choose between abridged versions on tape or CD, or drag along lots of tapes and CD's. Talk about convenience. You can look for it at Audible.com, and if you wish, sign up for a one- or two-book-per-month download. Audible can be used with: iPod, MuVo, several PDAs, and other devices. Go to www.audible.com to see what's currently compatible, then click the Device Center icon at the top of the Web page to see the list. For people who like to listen to books, this is one cool system.

Data Storage

Many media devices also permit you to store files from your computer, along with the music. This can be a handy way to ensure that a file is secure. After all, it's in your pocket, so you can defend it as vigorously as you would any other private items located near your pocket. Or you can use it to transfer work between the office and home, though I think it's simpler to just send files via broadband e-mail for this purpose.

You might have other reasons to carry data around. Some media devices also allow you to plug in memory cards of various kinds to supplement built-in memory. But the primary differences between simple storage devices and media players are that the former cost less and have few built-in software capabilities beyond the simple storage and retrieval of files.

Video, Photos

The high end for portable digital media includes PDAs (with their computer operating system user interface and capabilities) and media players that offer similar computeresque qualities but with a dedicated multimedia interface.

Media players—which currently cost about the same as a top-quality PDA ($400–500)—specialize in audio, of course, but add facilities for photo and video display. iRiver's PMP-120 features a 3.5 in. color display and 20GB of memory for a $449 street price. Creative's Zen Portable Media Center at $499 street offers the same storage and supports WMV, WMA, MP3, and JPEG formats on its 3.8 in. backlit LCD screen.

On the high end, Archos's AV480 pocket video recorder has a generous 80 GB and can record loads of TV programs via its cradle connection ($699). The unit's software also allows you to schedule multiple recordings and either watch the video on the included 3.8 in. LCD screen, or play it back on a TV. The manufacturer claims you can record up to 320 hours of TV (MPEG-4 SP compression, ADPCM stereo sound, in the AVI format) by plugging in most video sources (VCR, cable, or satellite). It can store and display up to 800,000 JPEG photos, and it also supports the Microsoft BMP photo format. Interestingly, the video playback is capable of DivX 4 or 5, an increasingly popular video compression technology. Playback also features MPEG-4 SP and MP3 or ADPCM stereo.

THE BUYER'S CHECKLIST

If you're considering one of these portable media players, use the following checklist to ensure that you're not forgetting a feature that's important to you:

- What is the capacity in gigabytes? (Video eats memory far faster than music or JPEG photos.)
- How long do batteries last? (Check both audio and video battery-use specifications. Video playback typically uses up battery power about three times faster than audio.)
- If the unit has a rechargeable battery, does it need to be replaced after a certain number of charges, or after so many months? (What's the cost of replacement? Is replacement easy, or do you have to send it back to the factory; or is replacement impossible and you have to junk the unit?)
- How about the LCD screen? (Is it big enough for you? Does it have the desirable backlit TFT LCD display? Do you need color?)
- What is the resolution of the display? (A resolution of 320 x 240 pixels is typical.)
- What video formats can it handle, if any? (DivX, ASF, DVR-MS, WMV, MPEG, MPG, MPE, M1V, MP2V, MPEG-2, AVI—you don't need them all, but ensure you get the formats you're interested in. For example, do you have a large collection of home videos on your hard drive stored in WMV?)
- What audio formats can it handle? (At a minimum, you want MP3 and WMA.)
- What photo formats does it use? (JPEG is generally the best compromise between file size and compression quality.)
- Does it permit video recording?
- Does it have an FM tuner, FM recording capacity?
- Does it support voice recording?

- What are its data storage facilities?
- Does it have a built-in speaker, or merely earphone or headphone output? Tip: Really comfortable, high-quality earbuds can cost you extra. Sony sells a good line of earbuds starting at around $40 and going up. Noise-canceling versions are $100–150, if that's an important feature to you (do you fly a lot?).
- Does the device have an expansion slot? (Can you add more memory or features such as an optional FM tuner?)
- What type of computer connection does it have? (USB 2.0 is best—the most common high-speed interface.)
- What A/V inputs and outputs does it have? (If you want to save your stored videos to a DVD recorder or display them on a TV in your motel room, you need audio and video line out jacks. Do you expect to play audio through outboard speakers or a stereo system?)
- What color is it? (Should you buy a unit that matches your accessories, shoes, or car? Do you care? White goes with everything, and the iPod's minimalist design is awfully stylish. Lots of people choose it because it looks so cool in the ads, white wires flying around gyrating beauties.)

4 RIP AND BURN

- ◆ How to copy, rip, and burn music CDs
- ◆ Audio quality compared
- ◆ Converting songs across formats
- ◆ Recording streaming audio
- ◆ Copying iPod songs to your computer
- ◆ Labeling CDs

You doubtless have the software you need to copy CDs. For the PC: Roxio's Easy CD Creator (or its new Easy Media Creator 7) and, my preference, Nero Express are bundled with many CD and DVD writers. For Mac there's Roxio's Toast. And of course almost everyone has the venerable Windows Media Player (see chapter 3 for details on how to use it to rip and burn). Any of these utilities are fine for most general-purpose custom CD creation.

The process is pretty much the same in all cases: you put a blank CD in your writer drive, select and specify the order of the songs you want to compile onto your custom CD, then click the Burn button. You sit back and watch the progress bars in the Rip/Burn utility as the music flows onto your new CD.

Options and details of course vary. So let's take a brief walkthrough using one of the best and most popular burners, Nero Express.

Savvy Tip

Burning is easiest if you have two drives: a playback-only drive and a recorder drive. That way, if you're copying some home movies on DVD or some of your personal musical creations from one CD to another, it's just a matter of putting the original in the playback-only (read) drive and the blank disc in the recorder (write) drive. You thus skip the step of having to swap discs after ripping from the original to memory, then later writing from memory back to the same drive. This is particularly useful when copying music on slower drives or when copying those huge movie files—you can start the process and go to bed. The copy is ready when you wake up.

Nero Express is simple, elegant, and works perfectly. As you can see in figure 4-1, the application gives you several options, including a bootable data backup disc; video (SVCD or video CD); direct bit-for-bit copy of an existing CD; and, for music stored on your hard drive (or another CD), audio (.wav files for example), WMA, MP3, or music plus ordinary data files (such as photos).

Note: Neither the publisher nor the author condones copyright infringement of any kind. If you are unsure as to the legality of your activities, visit www.copyright.gov or www.mpa.org/copyright/copy resc.html for details on copyright issues.

COPYING AN ENTIRE CD

To make a copy of an existing CD (containing a recording, for example, of your band's music), the process is extremely simple (you end up with a music CD playable on all CD players, not an MP3 or otherwise compressed version):

1. Put the original in the playback (source) drive.

2. Put the blank CD in the recorder (destination) drive. (If you have only a single playback/recorder drive, you put the original CD in, then later swap it for the blank after the tracks have been ripped.)

3. Click the Burn button.

CREATING AN MP3 OR WMA COMPILATION

To make a compilation in MP3 format of tracks from music files on your computer or on a CD, follow these steps (you end up with a CD that holds songs stored in the MP3 format):

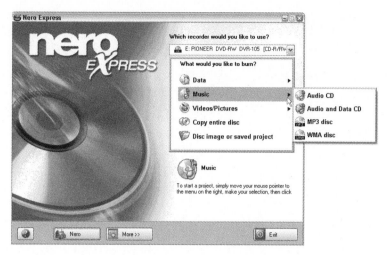

Fig. 4-1: Good burn utilities offer lots of options, from data to video to various kinds of music.

1. Follow steps 1 and 2 above.

2. Choose MP3 Disc from the pop-out menu shown in figure 4-1. You now see the dialog shown in figure 4-2.

3. Either click the Add button (which opens a file browser) or just use Windows Explorer to drag and drop whatever .wma files you want to include in your new CD.

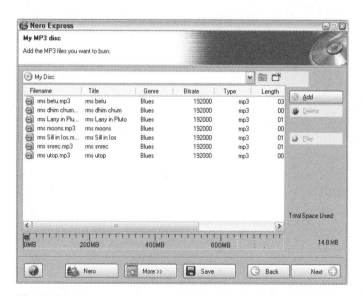

Fig. 4-2: Here's where you drop whatever MP3 audio files you want to copy.

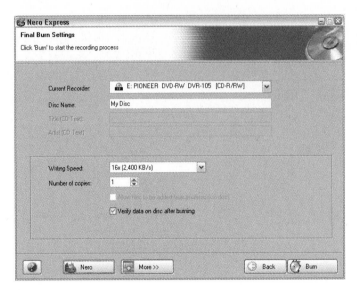

Fig. 4-3: Specify preferences for your burn in this dialog box.

4. Watch as the Total Space Used bar along the bottom of this dialog box shows your files filling up the disk.

5. When you're satisfied with the tracks you plan to copy to the new CD, just click the Next button.

6. You see the Final Burn Settings page shown in figure 4-3. Here you can specify if you want to take the extra time to verify the copy, make multiple copies, change the write speed, title the disc, and choose a different destination drive. If all seems as you want it, click the Burn button.

7. The progress of the burn is illustrated by the progress bar shown in figure 4-4.

Now you've got a CD that contains MP3 files, the advantage being that you can stuff lots more music using this format than ordinary uncompressed music CDs hold. If you want to double the amount stored by MP3, store your music as WMA files instead. Of course, fewer players support WMA than the wildly popular, and likely the default standard for years to come, MP3.

CONVERSIONS

For most of us the time comes when we want to transform some MP3 or WMA files to regular audio, so we can send a CD to grandma that she can play on her CD player (only the latest CD players rec-

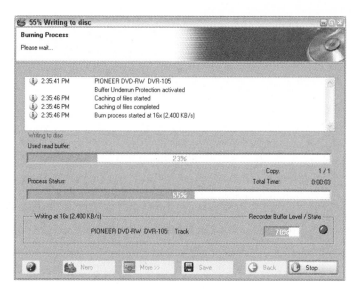

Fig. 4-4: You can see the burn, then the optional verification, in this status window.

ognize compressed formats like MP3). Or you might need to convert regular audio CD tracks to compressed formats like WMA or MP3 or whatever. What you need is an audio file format converter utility. Format conversion is technically called *transcoding* because you're converting one compression/decompression, or *codec*, to another.

Convert Regular CD Tracks to WMA or MP3

There are many audio format converters available. Most music applications do at least some conversions. For example, if you rip tracks from a regular CD using Windows Media Player, you can specify whether they should be converted to various WMA formats or a simple version of MP3. In Windows Media Player 10, choose Tools|Options|Rip Music to specify the format and sound quality among other options (see chapter 3 for details).

dbPowerAMP

For a useful, inexpensive all-in-one converter, you might want to give dbPowerAMP Music Converter a try: www.dbpoweramp.com. It's guaranteed not to infest your computer with popups, nagware, spyware, and other unwelcome annoyances that increasingly ride piggyback on shareware these days, and even in some famous commercial audio applications. You get to try dpPowerAMP free for 30 days, then pay a $14 registration fee, after which you get permanent access to MP3, DSP effects, and quite a few other features.

You can rip from audio CDs and store in a multitude of codec add-ins (you choose which formats you want, and you can download them free anytime). This application's conversion capabilities are particularly impressive: it offers every format I've ever heard of and some I hadn't heard of. In addition, their "Codec Central" download Web page is continually being updated as codecs are improved. You can work with all flavors of MP3, MP4, Ogg Vorbis, WMA (various), AAC, Monkeys Audio (considered one of the best), Musepack, FLAC (lossless), and others, even esoterica like Creative Labs Nomad Voice. Check out the available codecs, and their descriptions and uses, at http://www.dbpoweramp.com/codec-central.htm.

Conversions preserve ID tags (and you can edit them as well), and the converter includes a variety of useful features, such as volume normalization (so some tracks aren't recorded louder than others), digital signal processing (equalization [EQ], dithering, fade, and others), voice removal (to create karaoke files you can sing with), and *music removal* (so you're left with only the voice, and can do a reverse karaoke, like providing a piano accompaniment for your favorite songbird). But be warned: These latter two features offer mixed results at best and depend on qualities in the original recording. For example, voice removal works best if the singer is recorded in the center of the stereo field.

Here's how dbPowerAMP works. The default codecs are shown in figure 4-5, which shows the dialogue box that appears when you start the converter. They are: CD Audio (.cda), Wave Audio File (.wav), and MPEG's various flavors (.mpg, .mp1, .mpa, .mpga, .mp2, .mp3, .mpx). However, you can easily add any other codecs to these defaults via the Codec Central described above.

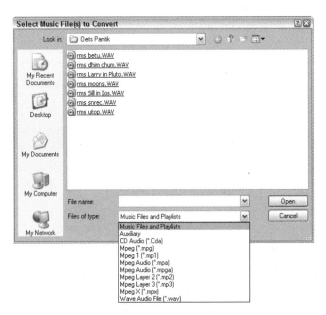

Fig. 4-5: dbPowerAMP allows you to choose from the default list of codecs shown in this drop-down list.

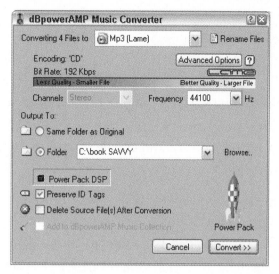

Fig. 4-6: Here you can specify the codec and lots of other options for your conversion.

Alternatively, you can just use Windows Explorer, to which the converter automatically attaches itself during setup. Select one or a batch of audio files, then right-click and choose the Convert To option on the context menu. Either way, after you've chosen audio files to convert, you see the main utility, as shown in figure 4-6.

Figure 4-7 shows you the wide range of conversion quality presets you can choose from.

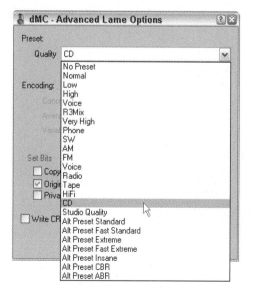

Fig. 4-7: Select from this generous list of settings to specify conversion quality.

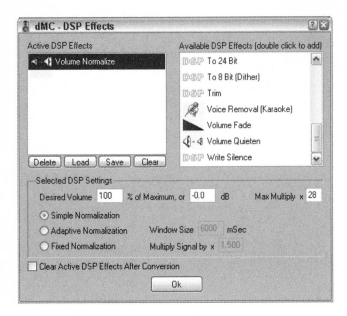

Fig. 4-8: Looping, trimming, reversing, adding silence, and other special effects are specified in this dialog box.

Your choice affects both the speed of the conversion process (Who cares? Where's the fire?) and the ultimate size of the resulting converted file.

Finally, click the Power Pack DSP button to apply any special effects during the conversion process, as shown in figure 4-8.

Those who do a lot of track management will probably enjoy the pop-out track specs feature that dbPowerAMP adds to Windows. Select a track in Windows Explorer and lots of details about the recording are displayed, as shown in figure 4-9.

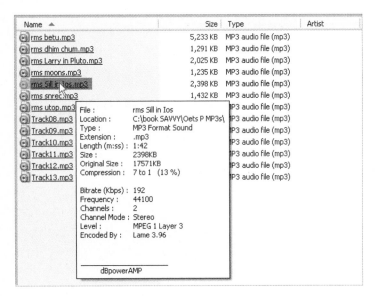

Fig. 4-9: Details about each track are a click away in Windows Explorer.

PRIORITY ADJUSTMENTS IN DBPOWERAMP

One nice feature that dbPowerAmp offers that I wish were more common in Windows programs is a priority adjustment (how much of the CPU's time the converter gets to hog). When you begin the conversion process, you can drop a listbox and choose from the following options: Pause, Idle, Low, Below Normal, Normal, Above Normal, and High. If you choose High, for example, the utility really takes over the computer during the process—moving the mouse has no effect except occasional wide jumps across the screen. But set it to Above Normal and you get rapid conversion with some limited multitasking permitted. The mouse, anyway, moves OK. However, you can make an adjustment in Windows to improve audio processing speed, boost the number of simultaneous effects you can apply, reduce latency, and so on. Most digital signal processing and other audio manipulation is carried out in the *background.* By default, the Windows setting makes background processing secondary to such foreground activities as application screen redraw, window-closing, menu-dropping animations, and other unnecessary business.

If you want to give your digital audio processing a real boost in power, follow these steps:

1. Choose Start|Control Panel

2. Double-click the System icon in Control Panel.

3. Click the Advanced tab.

4. Click the Performance Settings button.

5. Click the Advanced tab.

6. Click the Background Services option button.

BEST-QUALITY DOWNLOAD BURNS

Often, to get the highest quality audio from music downloaded from one of the online services, just use the direct-record-to-CD option. You can always later convert this quality format to a WMA or MP3 format using a conversion utility like dbPowerAMP. Set the conversion to a high bit rate such as 192Kpbs for your conversion.

A free, open-source application called Audacity also features a "what you hear" recording capability. Plus, it's a powerful audio (.wav) file editing and recording package as well. Take a look at it at http://audacity.sourceforge.net. Of course, you must only record audio that is not copyrighted unless you get permission from the copyright holder.

RECORDING STREAMING AUDIO

Internet radio, online music stores, streaming video—audio comes into your computer from the Internet in many streams, and also from other sources such as computer games. Can you save this audio to a file on your hard drive? Sure. The quality varies from wretched to wonderful, and if you're expecting top-quality streaming audio from a Web site, you do need a high-speed connection.

Recording Whatever You Hear

There are two primary ways to record audio that streams in from the Internet. The easiest is to buy a quality sound card that features a "what you hear" input. In other words, in addition to usual Line In, Synth, Aux, CD, and other recording input sources, there's one called *what you hear*, meaning that it records *whatever* sounds you hear through your speakers. In effect, the audio card's 7.1-channel analog speaker outputs are captured, so you can even record high-quality surround sound (in multichannel WAV format, so be prepared for big files).

If you're interested in this capability, ensure that the sound card you're considering buying has the ability to record whatever is heard through the system. One of the sound cards I use has a very long name: Creative Sound Blaster Audigy 2 ZS Platinum Pro. It comes with lots of support software, including the Creative MediaSource Player. Figure 4-10 displays the What U Hear input option. You can record up to the maximum quality of 24-bit/96 kHz.

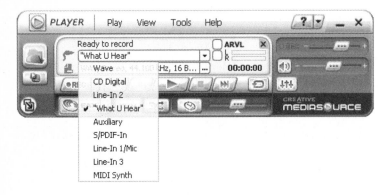

Fig. 4-10: Streaming audio can be recorded using Creative's MediaSource Player software.

Fig. 4-11: Save your recorded stream with the file name you specify here.

With the input set to What U Hear, just click the REC button on the software, and the recording begins. When you're done recording the stream, click the stop button as usual, and you then see the dialogue box shown in figure 4-11, where you can specify the filename and path.

Using Windows Media Encoder to Record Streams

This technique is a bit more complicated than simply recording what you hear, but the result is essentially the same: you capture streaming audio to a disk file. In this case, you can select CD-quality recordings that end up as WMA files (of course, you can always convert them to your format of choice using the techniques described earlier in this chapter).

Here's how to record what you hear via Windows Media Encoder.

1. Look for Windows Media Encoder in your Start|All Programs list under Windows Media. If you don't find it, download it from www.microsoft.com/windows/windowsmedia/9series/encoder/default.aspx.

Because the what-you-hear input records everything, be careful that you don't try to use the mouse (clicks are by default audible) or otherwise introduce extraneous noises into the recording.

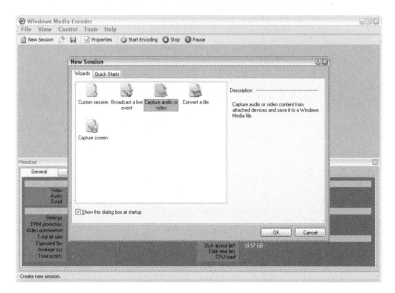

Fig. 4-12: Begin your capture by selecting this option.

2. Go to your online streaming site, or otherwise prepare the audio to be recorded, and press the Pause button on the site's playback interface or otherwise pause the audio at the start of the track you intend to record.

3. Run Windows Media Encoder, and you see the user interface displayed in figure 4-12.

   If you don't see the New Session dialogue box shown in figure 4-12, click the New Session option displayed beneath the File menu.

4. Double-click Capture Audio or Video icon shown in figure 4-12. You then see a New Session Wizard dialog box where you can select the device you want to use.

5. Select your sound card.

6. Click Next.

7. Type in the filename you want to use to save the audio. The .wma will be appended automatically for you.

8. Click Next.

9. You now choose the intended target (streaming, Pocket PC, server, and so on). Click the File archive option.

10. Click Next. Here you choose the recording quality, from sorry FM quality all the way up to lossless variable bit rate WMA, as shown in figure 4-13.

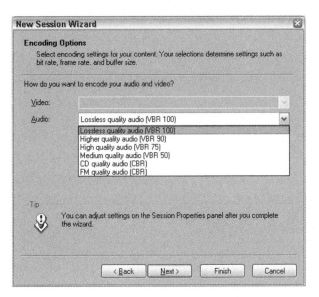

Fig. 4-13: Select the recording quality here.

11. Click Next and you see the specifications of your intended recording. Click the Back button if you want to make any changes to these settings.

12. Click Finish.

13. Click the green Start Encoding button to begin the recording.

14. Click Play on the Web page or other source to start the audio. You see the ongoing recording details displayed in figure 4-14. If you're getting up into the red zone at the top of the VU (volume unit) meters on the left side, stop the recording by clicking the red Stop button shown in figure 4-14. Turn down your system's volume control (it should be available via double-click on its icon in the system tray). Then restart the recording process to avoid clipping.

If you're only experienced with analog recording, such as recording to cassette tape, you need to adjust your way of dealing with the VU meter. It could be desirable in analog recording to briefly hit the red zone in the input (or "level") meter. Allowing red to flicker a bit from time to time indicated that you were probably at the right spot between distortion (all red all the time) and too much noise (no red anytime). Recording *digital* audio is different. It's OK to be in the yellow zone, but as a general rule you don't want to see the red at all.

However, meters differ, so you need to experiment a bit with your digital recorder's input meter to find the best recording settings. Some digital meters are more forgiving than others and allow you a

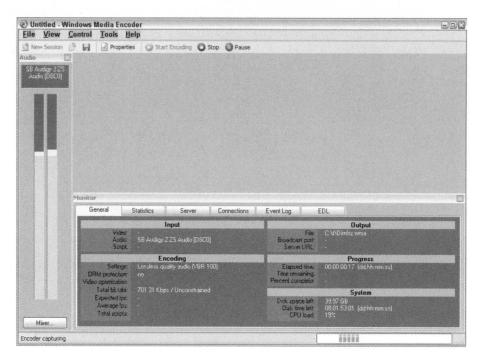

Fig. 4-14: Keep an eye on the level meters on the left side. If you're seeing red in the bars, stop the recording and lower the volume.

little flash in the red zone now and then. Remember the balance you're trying to achieve when recording digitally: the higher you get on the meter, the more data about the music you're capturing (and the less noise), but go too high and you clip off the sound waves, causing an ugly distortion in the sound. Examine the waveforms of the three recordings shown in figure 4-15. The one on the top is recorded at too high a level, too hot and loud. You can see that the tops and bottoms of the waves are squared off—they've gone beyond the recorder's ability to capture them. They've been *clipped*. The distortion isn't the same sound as overloaded audiotape, but it's unpleasant. I can't describe it. You'll know it, though, if you record it.

The recording shown in the middle in figure 4-15 is just about right: the sound waves are captured in their entirety, not clipped. The track on the bottom suffers from being recorded at too low a level. Here the signal-to-noise ratio will suffer because there's just not that much signal, and other aspects of sound quality are also compromised because you're not taking advantage of the recorder's capabilities. Boosting this signal later is a real compromise, just as trying to add contrast to a poorly shot photo never achieves the quality you'd get if the contrast were shot correctly in the first place. But remember that input meters differ, so you need to experiment to find the sweet spot on your meters—to determine whether hitting a little red (or even too much yellow) signifies clipping.

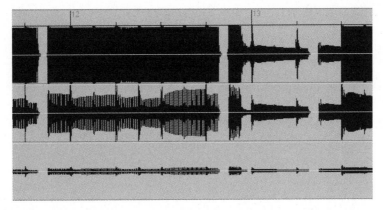

Fig. 4-15: Here are three results of recording the same music. On the top the input level was set too high, so the waves are clipped off on the top and bottom. The recording on the bottom is noisy because the input signal was too weak. The waveform in the middle is about right.

STORING iPOD SONGS AS COMPUTER AUDIO FILES

Here's the scenario: You've been a good iPod user and have bought lots of songs, but then your little sister accidentally erases most of them from your hard drive and you want to copy those songs from your iPod back into your iTunes library in your computer. You try copying them from the iPod, but that's not possible. What to do?

As is so often the case, the Internet comes to the rescue; you can download utilities that do the job of restoring your song library. Search Google for any of the following utilities: Pod Manager, iPodRip, PodMaster, or iPod Access for Windows, among others.

LABELING CDS

Many CD-burning utility suites and other software include the ability to print custom CD or DVD labels. Some experts' advice on this topic is *don't*. They point out that when CDs or DVDs are flexed (such as when you bend them to pry them out of their case), you're weakening the glued layers that make up the media. Likewise, normal temperature changes will slightly adjust the shape of optical media, and if you paste on a label, you create additional stress on the plastic strata. These experts recommend just using a soft felt-tip pen to write on the disc.

Savvy Tip

If you decide to label your discs, it's probably best to affix the label before you do the recording. This is particularly true for DVDs, which in some cases can take hours to record. Why? Because it's not that hard to mess up and get the label a bit off center, so the holes don't line up and the disc can be unplayable. You don't want to have to rerecord the whole video after messing up the label job. Also, don't try to remove an affixed label—you can really stress the CD's plastic sandwich and the reflective media layer it is designed to protect. However, some new CD burners feature nondestructive labeling such as etching. If this is an important issue for you, search the Internet for the latest CD-labeling technologies.

The other side of this issue is that aside from dual-layer DVDs ($10 each currently), the cost of most optical media is so low that you can simply create two copies, one that you affix a sticky label to, and the other for backup just in case.

Editing ID Tags

MP3 files include tags—data fields—that can identify the album title, artist, year, and other information. These details are displayed on portable music players, in Windows Explorer, and so on. If you want to edit this data, take a look at www.id3-tagit.de/english/index.htm,

id3-TagIT is free, and it has some nice features. For example, you can edit multiple tags at once (to, say, change the album-name spelling of all the tracks on that album) and convert filenames.

5 FACING THE MUSIC: "FREE" AUDIO, LEGAL AND SECURITY ISSUES

◆ What's copyrighted? What's free?

◆ Napster transforms, BitTorrent arises

◆ Free versus pay sites

◆ Peer to peer

◆ Avoiding spyware and viruses

◆ Free Internet radio

◆ Podcasting: time-shifting for audio

I'm not a lawyer, so the sections in this chapter discussing the current copyright mess are merely my opinions about a sad, sad situation, and it's getting more and more absurd.

Oops! I seem to have strayed into someone's copyright, and as I understand it I can write the words *sad, sad situation* and the rest *if* I mention that Bernie Taupin, Elton John, and who knows who else hold the copyright to those phrases.

Or perhaps not. There seems to be some dispute about many issues surrounding copyright. If you quote *too much* of something, you've done wrong and have moved into plagiarism. Or if you fail to put quotes around, or properly attribute (footnote) something, even something very brief, if it's somebody's *original idea,* you're bad. Now if it's an *unoriginal idea,* it's OK to copy it.

As I understand it, Wal-Mart has trademarked or copyrighted the word *always*! I suppose I can still use that word though. Always, always, always! There! Let's see if I get a phone call about it. But could I use "Always" in a sign at my business, or in an ad campaign? I guess not.

This is my 37th book, and I sometimes see my work for sale in used bookstores and even loaned out for free by libraries (I get no royalties in either case). In fact, libraries and used bookstores are repositories of even newer media such as VHS tapes, CDs, and DVDs, which can, evidently, be freely passed around or even resold, as long as they're not *copied*—except for tapes, which are of such poor quality that they allegedly *can* be legally copied *if you first buy one*. But you can't turn around and sell the copy, or even show it to others for a fee. I guess you could make a VHS copy of a TV show and give it to your mother, but if she then handed you a cookie—who knows? It really is getting more and more absurd.

WHERE'S THE USED SOFTWARE?

But you rarely see *software* for sale or rent. Evidently there's a distinction between a $13 CD or DVD and a $150 software package. I asked one popular local used bookstore why they stopped selling used software. They said that they didn't make much money on it and that they also had a real problem getting some of the staff to understand that some expensive software is useless without the registration necessary to license it.

Can you download music from the Internet without paying for it? Apparently if it's not copyrighted you can, otherwise no. Downloading is a form of copying. But some artists allow you to try their music free via downloading. It's a matter of permissions on various levels, it seems.

Perhaps you remember the fuss surrounding the introduction of consumer videocassette technology in the 1970s. Some movie studios claimed that allowing people to copy movies would destroy their industry (of course this claim has been made repeatedly through history, by different industries, such as when TV first appeared and the movie people made the same claim). Didn't happen. Never does.

The dust settled about consumer videotaping when the Supreme Court issued its famous "Betamax" ruling (http://caselaw.lp.findlaw.com/scripts/getcase.pl?court=us&vol=464&invol=417). This ruling stated, among other things, "Any individual may reproduce a copyrighted work for a 'fair use'; the copyright owner does not possess the exclusive right to such a use." Many interpretations have been made of this statement, among which is the claim that if you buy a DVD you should be able to make an archival copy of it. Challenges to that interpretation continue today.

And copy protections are being built into televisions, particularly HDTVs, where a digital stream can potentially be captured and recorded (instead of plugging the HDTV source—such as your cable box—into the TV, you instead plug it into an HDTV recorder, for example). The distinction claimed here is that unlike analog signals (such as those recorded by VHS), digital signals don't degrade from copy to copy. Digital is inherently different from analog—the number 1287, for example, remains that same number no matter how many times you copy it. It doesn't start to fade, or become hissy, or suffer from any other classic signal-loss issues. So digital transmission from, say, an HDTV cable box and an HDTV is being encrypted or something. They don't want you tapping into that stream of bits and recording a perfect copy.

NAPSTER HITS THE WALL, BITTORRENT ARISES

As for music, CD copying is common, and Napster grew huge for a time, apparently on the basis of the theory that if you *distributed* the storage of copies rather than storing them in a central repository, you somehow avoided copyright issues. Other ideas are thrown around: by *sharing* you're not actually *selling* (the old "library defense"). Lately a new twist has appeared in the form of Bram Cohen's clever BitTorrent software. Where Napster fractured libraries, Bit Torrent splinters individual files.

You do see the trend—each new technology further deconstructs the original, only to reconstruct it at the destination. It's getting rather close to Star Trek's beaming: transportation via atomization and restoration.

I believe that Mr. Cohen isn't interested in piracy but rather in improving the efficiency of moving bits from here to there—which he seems to have accomplished. When, however, a technology improves efficiency (just as videotape made time-shifting efficient for the average viewer), there can be side effects. Some people want to know if you can copyright what amounts to a mist of bits flowing in from all directions and becoming, finally, coherent on your hard drive. I suspect the answer from the movie industry is the same as always: We'll try to stop it. Some sites employing BitTorrent have been shut down.

How does BitTorrent work? Classic file sharing can slow things down because if 2,000 people are downloading a file from someone's hard drive, the process becomes slower for everyone. But if only one person is downloading, it's fast. With BitTorrent, though, the minute a bit is downloaded to someone's computer, that bit becomes immediately available for uploading to other computers.

Put another way, if 2,000 people are downloading a file from a single computer, and you start to download that file, you're actually downloading simultaneously from 2,000 computers, not just one. A bit here, a bit there, adds up to a torrent. And it's not sequential. It's a spray of bits, and, strangely, the more people who are downloading the file at one time, *the faster the process becomes for everyone.*

CAN I BE SUED?

Yes. As we all know, you can be sued for smiling at the wrong time.

And currently the antipiracy forces are filing lawsuits against individuals who share files on the Internet. Although usually not being sold, these files *are* being copied—even if via a kind of teleportation.

Digital and intellectual rights management is a complicated issue, and there are compelling arguments on both sides. After all, I do want to be compensated for writing my books, so I would prefer that people not copy them and sell the copies. On the other hand, I'm a big fan of used bookstores, so I can accept the lower income that results from this aftermarket.

However, given that *perfect* copies can now be made, content providers such as artists and record companies have an understandable fear that the new technology will seriously impact their livelihood. Is the downturn in CD sales revenue due to the economy in general, or is it peer-to-peer music "sharing"?

FREE VERSUS PAY SITES

Several online sites offer *legal* free downloads from artists hoping to make it, plus a few who *have* made it. You don't find Roxette or Green Day on these sites, but some of the music is terrific. And some of the folks contributing to free sites *are* famous, such as Conor Oberst and Bright Eyes. It's usually amateur night, true enough, but some of these amateurs deserve to move on up to pro status. One popular site where you can find tasty free samples is Cnet's http://music.download.com. You can use the music freely for personal or educational (noncommercial) purposes. Artists pay nothing to contribute their music to this site, and you pay nothing to download it. At the current time, auditioning via streaming isn't yet supported, nor are playlists, but these features are planned for the future.

The famous pay sites such as iTunes and Napster are examined in the previous chapter, but they generally follow the same formula: 99 cents a song. Although Napster has recently announced a plan permitting unlimited "refills" of your MP3 player for $15 per month. See chapter 4 for details on this interesting development.

PEER-TO-PEER SITES

And of course there are the equally famous "free" sites that employ peer-to-peer (P2P) engines to pass music around. Some claim to be entirely legal "to own" (meaning their software itself isn't illegal); however, people who actually use that software to distribute copyrighted music have in some cases been successfully sued.

Both the Recording Industry Association of America (RIAA) and the Motion Picture Association of America (MPAA) have taken legal action against peer-to-peer downloading. Recently the MPAA claimed to have shut down a BitTorrent-based network called LokiTorrent.

The DMCA

Adding to the confusion is the Digital Millennium Copyright Act (DMCA) passed by Congress in 1998, to expand copyright law into the new technologies brought about by the digital media and the Internet. This act was opposed by many librarians, academics, consumers, and scientists. It was supported by music, movie, and computer software interests. Among other measures, the DMCA requires webcasters to pay licensing fees to record companies, and prohibits selling or even just distributing technology that cracks codes in order to illegally copy software.

Watch Out for Malware

Some downloadable software, including some peer-to-peer applications, is notorious for installing spyware and other noxious malware during setup. If you're contemplating downloading *any* new shareware or freeware software, my best advice is to download from a reputable site like download.com. And before you actually install the software, *read online user reviews* about it. Users aren't hesitant to point out when popups, adware, or other exploitive junk has messed up their machine; they're happy to warn you off before you, too, suffer an invasion.

Also, online reviews sometimes point you to superior software that does what you're interested in doing, but won't burden you with nasty or downright evil consequences. Of course, you should read several user reviews to see if there's a consensus. One irate 12-year-old's opinion shouldn't be the basis for your decision. But if enough people have been using the scientific term *sucks* . . . well . . . I'd sure think twice.

FREE RADIO FREE AMERICA

How about Internet radio? Can't you set up a radio "station" on your Web site and broadcast music, or even your views that everyone is doubtless breathlessly waiting to hear? Sure you can, but you're still responsible for paying fees and getting permission to broadcast copyrighted music, no?

Mercora

One interesting service, Mercora P2P Radio, appears to get around this problem. It's described as P2P (peer-to-peer) radio, and given that listeners to your "station" cannot randomly access all your songs (they can select from only ten at a time) or download them—just *listen to your broadcast*—perhaps this violates no laws current or anticipated.

You broadcast your music collection and Mercora pays the music companies a fee for each streamed song. The service at www.mercora.com/ states its intentions thus: "Mercora is the largest music radio network in the world! We have combined peer-to-peer and internet streaming technologies to create the ultimate and legal music search and discovery service for you. With Mercora P2P Radio, you can search, find, and legally listen to hundreds of genres and thousands of artists in near-CD quality sound."

The idea is that you locate someone who broadcasts a "station" with taste in music similar to yours—then you just listen as they DJ the songs out to the world. It's a bit different from actually running a station. For example, Mercora randomly selects the sets of ten songs it locates on your hard drive, but you can govern this process by modifying the songs that you store or even creating playlists for fans of your broadcasts.

Also, a listener has the ability to randomly access songs, to a degree. You can search for songs, but a song must be currently broadcasting, not simply stored in someone's collection. So your chances of getting to listen to a classic Stones song at any given time can be iffy. When I last looked, the stats published on Mercora's site were: over 24 million songs in the network, 18,416 stations online, 6,355 artists available, in 390 genres.

Apparently "digital rights" restrictions limit you to no more than four songs per artist, or three songs per album, per hour. Some people will not have a problem with this kind of thing—enjoying for example a "shuffle" of songs in genres they enjoy. It's like listening to a favorite radio DJ who is in tune with your favorite sounds. For others, restrictions like this may be quite annoying.

Be on guard: You might find that downloading some proprietary (or even famous but annoying) software is required before you can listen to some stations. Fortunately, excellent stations like BBC start streaming without any fuss or fixes.

All Kinds of Free Radio

The Internet is loaded with radio stations, some traditional like BBC radio (www.bbc.co.uk) and others that are startups from everywhere. And if there's a broadband connection on your computer and theirs, the quality can be excellent (CD quality is not uncommon).

Take a look at www.radiofreeworld.com to see lists of free Internet radio stations. You're bound to find some that reflect your tastes and interests. Also see if you like www.live365.com, another gateway to lots of stations. They're listed by category and also by audio quality.

Fig. 5-1: Traditional broadcasters offer excellent online radio. The BBC, for example, has music for every taste I've ever heard of on their fantastic site, Asian flavas among them.

PODCASTING ANYONE?

Although you can podcast without an iPod, this twist on Internet music sharing (and personal talk shows) is named after the famous Apple device. Indeed, you can send out your program any way you want. Podcasting does for Internet music and talk shows what TiVo does for television: it allows time-shifting. A listener can receive your podcast for later listening during her commute into Boston. A podcast is an audio file, and in many cases it's also a way to "subscribe" to automatic downloads of future editions of your favorite podcast shows.

Listeners don't have to do anything special to get your show downloaded—it can be automatically sent to your subscribers. Subscribe and your machine ensures you get any new shows.

Every kind of talk show, far-out indie music from the Thai underground scene, in-depth cooking advice for those who love goat, rants, raves, whatever you like to listen to, you're likely to find a show just up your alley. You might think podcasting is audio blogging, but podcasting is not limited to indie shows; Disney, NPR, and others are sending out their podcasts, too.

Podcasting is also a way of making Internet radio portable, automatically. You wake up in the morning and the show is already stored on your portable device, ready for you to dump it in your purse as you hurry out the door. Or if you don't use a purse, dump it wherever. The *point* is that you're carrying a little Internet audio with you the easy way.

Shortwave aficionados for decades have used a trick to listen to their favorite programs from Radio Netherlands or Vilnius or whatever. They attach a cassette recorder to their radio, then set a timer. But for iPod users, the idea of time-shifting radio programs is entirely novel. Plus, it's easily the most automated kind of entertainment recording imaginable. After your initial setup, you don't have to do anything but remember to attach your device to your computer.

To get into podcasting, download iPodder from www.ipodder.org/directory/4/ipodderSoftware. You'll find versions for Windows, Pocket PC, Mac, Linux, Smart Phones, and other hardware.

Savvy Tip

If you're interested in hearing programs that have been broadcast on BBC or other online stations, many sites allow you to download programs for later listening. It's pretty much the same as getting podcasts, just not as automatic. You have to download them by hand, and then send them to your device by hand. But that's not too much trouble, is it?

A primary difficulty right now with podcasting is that it's in its infancy, so there's as yet no single, central clearinghouse, no "best" site you can go to and find links to the kind of "show" you're interested in hearing. However, give these a try:

- podcast.net
- Podcastalley.com
- Odeo.com
- www.apple.com/podcasting

Currently, podcasting (and podcast clearinghouses) are fairly free of advertising, but you can bet that will change once business realizes the potential. Talk about niche marketing and targeted audiences. How much more specific can you get than the audience that subscribes to the podcast *People Who Dress like Joan Crawford*? I can predict that if they advertise on this podcast, sales will skyrocket for the factory in Jersey that still makes shoulder pads.

6 MAKING THE MOST OF YOUR MP3 PLAYER

◆ Making quality recordings

◆ Hooking up your player to other stereo equipment

◆ Using add-on speakers

◆ Ensuring a high-quality signal (FM transmission, tape head coupling, direct plug-in)

◆ Wireless transmission

◆ Accessorizing with speakers, cables, waterproof cases, and more

This chapter is all about getting the best-quality sound, and the most flexibility, out of your digital audio device. You'll find out how to attach to car and other stereo systems, how to listen safely in the shower, how to store digital photos in your iPod, and other cool tricks. (I meant safely in the shower *for your portable player—you're* in no danger from those batteries.) But first, let's review the main points about getting good-quality recordings in the first place.

HOW TO MAKE GOOD RECORDINGS

If you record MP3 music off one of your CDs, what bit rate do you choose? Presumably you're making the copy because

a. you want an MP3 computer copy in case you lose or destroy the CD;

b. you're going to listen to the music on a portable MP3 device;

c. you plan to plug the portable media device into a stereo system; or

d. some other reason.

No matter whether you're archiving or planning to listen on the subway via cheap earbuds, it's always a good idea to assume that someday you might be playing that same music through a high-quality stereo in a quiet room. In other words, just because you can get away with low-fi on the subway doesn't mean you should record at that quality; you shouldn't assume that you'll always be listening to a particular song in the worst-case environment.

The best advice is to choose a variable bit rate, or VBR, to allow the recording to increase the rate of sampling as needed to ensure high quality. If you specify a minimum bit rate of 128Kbps and the "best quality" setting for your recording, you're likely to get a result that varies—as required—between 128 and 160Kbps or so. If you specify a constant bit rate of, say, 320Kbps, you're actually wasting storage space because that high a bit rate simply isn't necessary for a quality recording. And if you listen to lots of rap, for example, which is mostly spoken voice rather than actual singing, you could try lowering the floor from 128 to say 96Kbps minimum and see if you can tell the difference. I can't.

Try some experiments: Record the same song at 160Kbps *without* VBR (in other words, the constant bit rate setting), then at 320Kbps also at a constant bit rate. Compare them using a high-quality stereo system or high-quality earphones (I recommend Sony MDR-V6s, if you can find a pair). Do you hear a difference? Now, make one more recording of the same song with a variable bit rate and a minimum bit rate of 128Kbps. Again, see if you can hear the difference using the best music system you've got. I suspect that like most of us, you won't be able to tell the difference. And by ripping at 128 minimum plus VBR, you'll certainly get a lot more music on your portable storage device than if you specify some high constant bit rate.

GETTING HOOKED UP

You've got an iPod or some other cool portable music device. You've followed the advice in this book about making good-quality recordings. But how about plugging your portable unit into the stereo on your yacht or, more realistically perhaps, just plugging it into your home or car stereo system? Can it be done?

Yes. Radio Shack can help you find a cord that plugs into the earphone jack on your portable unit and into the input jacks on your living room receiver, car radio, or the fabulous media room in your boat.

Let's consider good/better/best hookup solutions. Interestingly, the lowest-quality connection is the most expensive, and the highest quality—a direct cord—is the least expensive. In this comparison, I'll talk about car stereo systems, but the same ideas apply to other transportation systems, such as luxury yachts.

Good Hookup: FM Transmission

You'll see ads for a device that plugs into your portable unit and broadcasts (well, perhaps "broad" is the wrong word for the very limited range of casting) an FM signal. Usually there are four or so frequencies to choose from, so that when you tune your radio to your iPod's "station," you can find one that presumably isn't already being used by a real FM station in the area. Radio Shack, Wal-Mart, and others sell these adapters for around $30.

The beauty contest winner is the white (what else) iPod cradle called Griffin RoadTrip ($79; you pay for gorgeous, just like in the real world). It plugs into your car's power or lighter, docks the iPod, and broadcasts on *any* FM frequency. However, you may find that it requires frequent adjusting while traveling, as interference can cause problems.

I found two problems when I tried one of the less expensive units, but I did only try one, so perhaps my experience wasn't representative. First, I live in a moderate-sized city, Greensboro, North Carolina. I was unable to find any interference-free transmission frequency. Second, and perhaps worse, even if you live in the middle of nowhere and have a clear frequency, you're still going to only get FM-quality, best case. FM is good enough in some situations. It's certainly fine for spoken audio books or for casual music listening. But after all, you've taken the time to make good variable bit rate recordings, haven't you? You want to hear that quality, and you can in many of today's better car stereo systems.

Better Hookup: Cassette Adapter

A step up from FM is a mechanical coupling device that looks like an audiocassette with a wire coming out of it. These adapters sell for around $20. You insert the "tape" into your car's cassette player and instead of playing an actual tape, the playback head picks up the signal flowing from your portable unit into the silvery contact on the adapter. A primary drawback to this solution is that it looks kind of weird to have a wire between your cassette player's drawer and your portable unit. It's the opposite of stylish. Unsightly. If you've paid for a great-looking iPod or Zen player, attaching a creepy black cord to it is rather a comedown.

Another drawback is the nature of the connection: it's better than FM quality, but it's not CD quality (and presumably your MP3 songs or other ripped audio is CD quality or better). So you get a compromised connection. It *is* a partly mechanical connection. Also, if you have a cheap cassette player, you're perhaps in for both mechanical and audio-quality problems beyond those I've experienced. I've used a cassette adapter for months with no mechanical problems and with acceptable audio quality for listening while driving. Listening while idling is another story. I opted for a high-quality car audio system, including JBL speakers driven by a sufficient amp, so when the car is just sitting there, without the swoosh of driving noise, I can hear the difference between the cassette connection and a CD. So I usually pop in a CD while waiting for a friend.

If you want iPod upscale, consider Dr. Bott's iPod Universal Connection Kit with Tape Adapter (around $50 street price). It's an all-in-one, all-iPod compatible kit: car power connector, cassette player adapter, a case for the iPod, stereo cables, and a FireWire PocketDock, which allows you to connect a standard six-pin FireWire cable to your iPod's docking port and charge it or swap files.

Best: Direct Hookup

Some newer car audio systems have an input jack. (Why don't they all? It would add, what, $1.50 to the cost of a car radio?) Direct plug-in is the best possible connection between your portable unit and the car's (presumably) fine stereo amp and speakers. The music goes right into the system without having to suffer the degradation of traversing the semi-mechanical connection of the cassette tape adapter or the interference and low-fi of FM transmission.

This is also the solution if you're hooking up to your home stereo system. What you want to ask for is a mini-to-RCA adapter cord. The line-out jack or headphone jack on your portable unit is a "mini" stereo jack, and you're connecting it on the other end of the wire to RCA "line-input" jacks on your stereo system. These cords cost less than $10.

Wireless speakers are an option, too. Hook them up in the house and carry the speakers to the deck, garage, or wherever while your iPod is on shuffle. Connect from headphone jack to the back of the car stereo using the CD changer jack. This is sometimes located in the armrest or on the back of the radio head unit. This approach requires wiring made for your application and can be purchased for around $75 from places like logjamelectronics.com. It's a clean, direct method, and the wires can of course be hidden.

USING ADD-ON SPEAKER SYSTEMS

How about making your portable unit a little less portable by carrying some quality speakers around with you? Well, perhaps that's not such a bad idea for picnics or for an office setup—kind of a compromise between a full stereo system and earphones. And, of course, with speakers you get to share the music with others.

There are various portable speaker systems around, and all offer the necessary built-in amplifier (a portable device cannot be directly connected to speakers—the sound needs a boost first). Some of these systems are truly portable (they run off batteries); others, like the Bose SoundDock, need to be plugged into an AC power source.

Here are a few add-on speaker systems for your consideration.

- *Altec Lansing's InMotion Speakers.* Altec Lansing makes InMotion Speakers for iPod ($149), though I'm certain that the speakers won't care if you plug in other MP3 devices. This unit boasts MaxxBass technology for what's described as "deep bass without a subwoofer" . . . we'll see about that. The specs also claim that four AA batteries offer up to 24 hours of continuous playback, 4 watts RMS power, response 60 Hz—20 kHz, and an S/N ratio at 1 kHz input of more than 75dB.
- *Bose's SoundDock for the iPod.* At $300 you'll expect some significant quality, and Bose has that reputation. As its name implies, the white-on-white SoundDock connects via the iPod dock connector, so it's *not* for other players. It includes a remote control and does accept all dockable iPods. It cannot run off batteries.
- Also take a look at three additional, well-regarded, portable speaker systems: the $80 Harman Multimedia JBL On Tour, Virgin Electronics's Boomtube for $90, and the Philips DGX320, $70.

TRANSMIT MP3 VIA WI-FI OR BLUETOOTH

Can you plug your MP3 portable audio device (or any other source of audio, such as a CD player or stereo system) into your home's wireless network? Sure. Several products are available from such familiar names as Linksys, D-Link, and others. Most employ Wi-Fi to send music around your house, though a couple use Bluetooth.

D-Link's DSM-910BT is a $125 kit that includes two small Bluetooth units. One connects to your MP3 player or other music source, and on the other end there's a receiver that plugs into your powered speakers. You can also purchase the transmitters or receivers separately, to connect more than one source or target.

Terratec's Noxon Audio unit (about $150) also broadcasts audio around the house, but it uses your Wi-Fi network rather than Bluetooth. It works on Windows (2000 or XP), Linux, and MacOS X

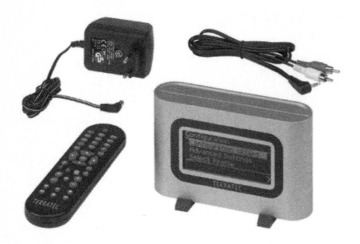

Fig. 6-1: The Noxon Audio uses Wi-Fi as its transmission method—wirelessly sending audio or Internet radio shows to any powered speakers or stereo systems around your house.

(including iTunes Contents). This unit includes a remote (most competitors don't), and it also has facilities for streaming Internet radio, *independent* of any participation from your computer. Again unlike most competing broadcasting gadgets, the Noxon has an LCD display, as you can see in figure 6-1.

Higher-end, whole-house music transmission systems include the sleek Sonus with loads of features and lots of flexibility. You do get what you pay for—some will say—but this system costs $1,200 for a starter system, and each additional Zone Player is $500, each additional controller is $400.

iPod users might want to look at the proprietary Apple Airport Express ($125). It works with the iPod system only, of course, but has some unusual features, such as permitting you to share a broadband Internet connection and USB printer.

And don't overlook the well-regarded Roku SoundBridge units (models range from $200 to $500). Among other things, it's one of the best-looking systems available, and it also handles Windows Media DRM or Real's streaming audio.

Savvy Tip

Before buying a home wireless music transmission system, make sure it can handle all of the types of streaming you plan to use. All units can broadcast music files located on your hard drive (though the Airport Express is limited to iPod format, and others cannot access iPod tunes). However, if you want to be able to transmit Internet radio stations' offerings or specific streaming systems (such as Windows Media DRM), check the specifications of the hardware before purchasing it to ensure that it does what you want it to do.

OTHER PLAYER ACCESSORIES

The enormous and continuing success of portable digital audio devices has generated a little galaxy of aftermarket products: convenience add-ons, bass boosters, equalizers, cases, shower-ready speaker systems, and other gadgets to extend the life, quality, or utility of your little precious.

Speakers are obviously the biggest boost you can provide—adding both sound quality (usually) and greater convenience (always) for those *moments musicaux* together with your special friend. (But if you don't want to spring for speakers, you might want to get Belkin's headphone splitter—in iPod white—that lets you share the lovely music by attaching two headphone sets to one digital music unit. At $3.49, the splitter costs much less than speakers.) And the two of you—so cute!—can also ask for two straws and share a milkshake while you're at it.

Let's take a brief tour of some of the more interesting smaller accessories.

Belkin makes lots of specialized accessories, including devices to use your player with a car stereo system. But they don't stop there. You'll find a holder that supports an iPod in a car, speaker and headphone splitter adapters, FireWire PCI cards, and other items.

Everything an iPod Could Ever Want

Need a special storage case for your iPod earbuds? Take a look at the earPod, a felt-lined case complete with a "rotating door" and belt clip, the removable "e-Clip." Interested in skins, extra batteries, links to your digital camera, cases, alternative earphones, voice recorders, car or wall chargers, tattoos? Just about any iPod accessory you can imagine is available from the iPod Accessories Store at www.audio-outfitters.com/index.html. According to the site, AudioOutfitters number one best-selling accessory is the earPod. To see the other items on their top-ten list, visit www.audio-outfitters.com/page4.html.

Another good source of information about all things iPod is www.ipodlounge.com/.

Equalization

The Koss eq50 is a powered three-band equalizer that you attach between your unit's headphone jack and your earphones. Two AAA batteries run it for 15 to 20 hours.

EarJams

If you're fond of bumpy, pumping bass (in other words, *extra* deep pumpers), try out the $15 Griffin EarJams. They attach directly to the iPod earbuds and both increase the overall volume and add additional bass. On the downside, this changes the sound, so some purists may find the "addition" is actually, for them, a subtraction. Give the EarJams a chance, though; it might be just the sound you like.

MP3 in the Shower

The dreamGEAR Boom Boom Multi-Box can be hung over the shower head (it has its own built-in hook). Or you could fearlessly take it to the beach. It works with iPod or any other audio player. You strap your player inside the cleverly designed plastic case, attach the music player to the Boom Boom, snap it shut, and you're off to the water park with no worries and a pretty decent-sounding set of speakers. An instant water-resistant boombox for those cookouts with your surfer friends. This is a really cool idea for only $20. Available in what dreamGEAR's Web site accurately describes as "six vibrant colors," it's big enough to hold some CD players. See the video demo for yourself at http://dreamgear.net/boomboombox.htm.

Downhill Racing with iPod

If you don't need to add speakers but want waterproofing nonetheless (for those hot-tub or skiing situations), seal your iPod off from damage with the $40 LiliPod from Eroch Studios: www.lilipods.com. "It's shock resistant. Watertight," says the Lilipod ad, and you need not worry about the stereo connector (it's watertight too), and the whole case is sealed off from outside leakage (or whatever might try to get into your iPod) with a compression clamp and an O-ring.

7 BUYING A SOUND CARD

◆ Understanding the creative marketplace

◆ Buying the right card for a gamer

◆ Audiophile cards

◆ Improving your portable's sound

◆ Sound card buyer's checklist

Two upgrades you can make to your computer will improve the quality of its audio: buy a better sound card and add good speakers. If you care about realistic game sounds or quality music, you'll want to seriously consider replacing the motherboard-based chipset, or el cheapo default card, that comes installed in most computers.

Let's be clear from the start: for most people, for *consumer* audio, today's manufacturers of choice for solid, reliable, quality, feature-rich sound cards include Creative and Creative. Or you can also consider Creative. Others have either fallen by the wayside (gone bankrupt or just thrown in the towel) or, if still competitive, aren't nearly as dominant in the marketplace. Like it or not, dominance has consequences. The Creative Sound Blaster line is the big gorilla of computer audio cards, and its popularity and high-volume sales have two effects. First, Creative's R&D budget tends to keep the company out front with the latest technologies and improvements (and other companies, such as game manufacturers, more readily jump on Creative's bandwagon, happily employing inventions

The cheapest sound cards are essentially mere signal pathways, sending audio here and there like a simple switcher, with little actual audio processing. Upgrading to a quality card can improve your computer's overall performance—particularly for multitasking and game play—by employing chips that take the burden of sound management off your computer's microprocessor. This unloading of CPU burdens can also improve latency problems—delays between your input and the response during music recording (discussed in "Defining Your Goals").

such as Creative's EAX technology). Second, Creative can sometimes undercut the competition by offering more quality for less cost. When you search for a sound card, ignore Creative's offerings at your peril. However, if you're after a sound card that's optimized for music (not for gaming), you probably want to also read up on audio interfaces from such manufacturers as M-Audio. One good source of advice can be found by checking with Sweetwater.com or zZounds.com. Also ensure that the upgrade card or interface you're considering is compatible with your software. For example, if you use Tascam's GigaStudio sampler, can you get a driver that makes it work effectively with the audio card or interface (I say *interface* because some audio peripherals are outboard, connecting via USB or even PCI card slots on portables)?

DEFINING YOUR GOALS

Before setting out to research the right sound card for your system, consider how you mainly employ your computer's audio capabilities. This section examines the uses to which sound cards are put and suggests the best solutions for two of the three primary categories of computer audio: games and home theater / music. Choosing a card for a computer that's at the center of a personal music studio—a card that can handle the demanding recording and mixing jobs—is more complicated. The options—including specialized semi-pro gear—are covered in chapter 13.

While the game player and music / home theater hardware categories are not entirely mutually exclusive, considering your primary use can help you refine your thinking about what features you really need. And a card optimized for games is, by definition, not optimized for audio applications.

AUDIO FOR GAMERS

If you're a game enthusiast, your card doesn't need to meet the demanding specifications of high-end audio. Leaping monsters and crashing cars—and the soundtracks that accompany most computer

Savvy Tip

Two technological improvements in the past few years in sound cards are now fairly common, and you might as well go for them when you upgrade your card. First, 3-D sound systems standards appeared (along with special effects sound processing—digital signal processing, or DSP—such as echo, reverb, and so on in some cases). Creative calls one of its primary features EAX, and this technology has been widely accepted by gaming manufacturers and is used in many of today's games. Other technologies include A3D and Directsound 3D. For game surround sound, look for the card's compatibility with Directsound 3D. The second and more recent improvement is 7.1 sound, though few commercial speaker systems are available that are capable of employing *that* much surround; instead, you can fall back to 5.1 for the time being.

games—are not the same as exquisite musical experiences and don't require the greatest stereo system.

However, if you're interested in spine-chilling effects, like the moans of the undead as they shuffle up behind you, you might want to consider going beyond stereo. There are several options. The least satisfactory is simulated surround sound (sometimes called simulated 3-D), but it *sometimes* works and the audio appears to come from behind now and then—if everything is just right. To achieve these effects, various psychoacoustic manipulations, such as adjusting the phasing, are employed to trick the ear of the psycho (that's you).

However, true, discrete 5.1- or even 7.1-channel audio, and the rear speakers to back it up, will send sound to the back of your ears. No tricks necessary.

RECOMMENDED CARDS FOR GAMERS

Good

Whatever card came with your PC might be good enough for games. If you mainly use your computer audio for car crashes, machine guns, and the like, the only necessary technical spec I know of is *loud*. And the burden of reproducing the best loud falls more on the speakers than on the sound card itself. So put your money into a set of *loud* speakers. But if you want refinements like EAX or surround sound, which are featured in many of today's games, read on.

Better

If you want your game soundtrack music to sound good, and you want true surround sound for those hair-raising special effects creeping up from behind, consider the reasonably priced Sound Blaster

Audigy 2 LS at around $60 or the 2 ZS / Gamer at around $80 (prices can vary widely over time). These cards have excellent audio specs and all the game-related audio features you're likely to want (plus some cool high-end audio features as well). If you want more flexibility—or don't like opening your computer to install new cards—check out the external, USB versions of the Sound Blaster cards; you can easily switch great sound between computers, and even take the show on the road with your portable.

Best

Want to go all out and get the top-rated consumer sound card that's also great for games? The winner in this category is the Sound Blaster Audigy Pro. There are two very similar models: You can go for either the 2ZS Platinum Pro or the 4 Pro. They're very much alike, but the 4 Pro costs about $275 street and the 2ZS Pro is $50 less. For gaming, and indeed most any use, the 2ZS is a high-end, powerful, great-sounding card. The 4 Pro features one additional stereo input (six inputs versus the five inputs on the 2ZS), some slight additional software (timed recording so you can set it like a VCR for unattended Webcast recording, though bundled software can change, so check this if it matters to you), improved DACs (digital to analog converters) and SNR (signal to noise ratio). Technically, the primary difference is that when recording 192 kHz / 24-bit stereo, the dynamic range for the 2ZS is 107 dB versus the 4 Pro's 113 dB. For me, and likely you, this difference is inaudible. In other words, for the recording studio purist with golden ears, the 4 Pro has a possible advantage. For gamers, the improvements are likely irrelevant. And for most of us, the differences cannot be heard anyway.

Both units feature external hardware—hubs that offer MIDI, digital, and analog I/O jacks, volume control, and other features. It's sometimes convenient to have knobs to turn rather than adjusting volume via software sliders onscreen. Likewise, the external hardware makes it easier to quickly change input and output cords. For about $50 less than the 2ZS Platinum Pro, take a look at the 2ZS Platinum model. It's similar but instead of the stand-alone box for I/O and knobs, you slide a hardware unit into a 5.25-inch drive bay on your computer.

Savvy Tip

Computers never rest, and sound cards will eventually need to make the transition from the now common PCI to PCIe (PCI Express) interface. PCI Express offers, guess what, faster throughput. But that's probably going to impact video cards sooner than audio, though you'll want to make the switch in the coming years. For now, if you want what I consider the best of the consumer audio cards, go with the PCI-based 2ZS or 4 Pro units. They offer quality sound to spare.

AUDIO CARDS FOR PLAYING SYNTHESIZERS

If you intend to play keyboards or other MIDI instruments connected to your computer, employing *soft synths* (sound banks or samples such as VST instruments held in the computer), you need a special kind of audio card designed to overcome the latency problem. Latency is when you press a note on your keyboard but there's a delay before that note sounds. This makes it difficult to record your music in real time as you play; for example, there can be a half-second wait between note presses and hearing the result. Special audio card drivers have been designed to overcome latency lag time. Such cards employ the low-latency ASIO (audio stream I/O) protocol, a special technology originated by Steinberg, the famous computer music house. Sound Blaster Audigy and Audigy 2 cards ship with an ASIO driver.

MUSIC AND HOME THEATER

If you're expanding your PC's capabilities into playing quality stereo, serving as part of a home theater system, you're not alone. The PC can add great features, such as automated recording, to an audio/video center.

Sound cards for stereo listening are essentially the same ones you'd want to use for home theater: great specifications, and perhaps special capabilities like DVD-Audio, which offers better-than-CD surround sound recordings. DVD-Audio was introduced in 2000 and there are several hundred discs available now.

For stereo music, though, surround sound is at this point optional, and true surround (discrete channels on a DVD-Audio as opposed to synthetic surround generated by a receiver) is quite rare.

You can add a stereo receiver near your computer or send sound via Wi-Fi or some other transmission method (see chapter 6 for some ideas) to a stereo system elsewhere in your home. In either case, an amplifier (in a receiver or powered speakers) is necessary for boosting the relatively weak signal coming out of the audio card.

Home theater and stereo listening don't require the demanding specs that you need if you're recording live audio, a topic covered in chapter 13.

Good

First, attach your stereo card to your stereo receiver or powered speakers. Listen for a few days and see if you are satisfied. Perhaps the sound system you bought when you ordered your PC in the first

> **Savvy Tip**
>
> The Audigy 2ZS model seems geared to home theater, offering the latest surround sound technologies such as DTS-ES and Dolby Digital EX. But this model also has plenty to offer those who only want to listen to music. The specs are good. And if you want to fiddle with the sound, the bundled MediaSource player/recorder (available with several Creative cards) includes lots of special processing effects you can pump the music through, such as Environment (EQ presets), Advanced EQ (presets for various kinds of music, such as heavy metal, techno, voice, and so on), flanger, tremolo, simulated 3-D from stereo sources, and so on.

place is plenty good enough for your needs. Listen especially for noise (hum, hiss, or other extra sounds, particularly during silent passages on a CD). The inside of a computer is often in a notoriously noisy environment, so the cheapest cards can suffer from noise leaking in past their shielding efforts. If this doesn't bother you, fine. After all, during playback there are often ambient noises that interfere with the music. Depending on where you live, these can include children playing, traffic noise, announcements from the warden, and so on.

Recording, though, is a more demanding situation; you don't want to include noise on your recording of a one-of-a-kind, perfect guitar riff. The rigors of recording via sound card are explored in chapter 13.

Better

What do you usually listen to? What's your source: Internet radio, streaming playlists from Napster, MP3 files, CDs in your optical drive? And what kind of speakers do you have? Little, cheap, bass-free satellites; boombox-quality thumpers; high-quality "computer" speaker systems; or a first-rate stereo system? Your answers to these two questions determine how good the sound card has to be. And, beyond those considerations, are you planning to upgrade the source audio or speakers anytime soon? You can find several midrange (in price and quality) options. Look at Creative's Live! 5.1 ($30), Audigy 2 ($70), and Audigy 2ZS ($90). With any of these devices you're likely to see an improvement over a default card that came free with your computer. Of course the more you pay, the more you get, and moving up to the Audigy line represents a step up in specifications.

Have you run out of slots to add new cards in your computer, or do you want an outboard card that works with your portable or can be added to your next computer? If so, there's a portable alternative: USB outboard sound hardware. Look at the $130 Audigy 2 NX model. It has great specifications in a cool-looking external box. Specs resemble those for the Platinum line.

Best

Simply the best consumer audiophile cards available are the 2ZS Platinum Pro and the 4 Pro. Read about them in the "Best" entry in the previous section. These units are the best for games, music, home theater, and amateur home recording as well. In other words, these cards do it all.

SOUND CARD PURCHASER'S CHECKLIST

When you've figured out what kind of person you are—gamer, music lover, music maker—and have narrowed down your purchase to a few cards, here's a final checklist to look over. See if you need (or think you may need *sometime* in the future) any or all of these features. When you've finished looking over this list, you should be able to find the one perfect card for your digital audio needs.

- Speaker channel output 7.1, 6.1, 5.1 or 5.2 (stereo)
- Digital audio coaxial S/PDIF (Sony/Philips Digital Interface, a standard audio transfer file format) in/out jacks, like those found on DVD players, newer audio receivers, and stereo equipment
- Optical S/PDIF in/out jacks like those found on DVD players, newer audio receivers, and stereo equipment
- 1/8 inch digital out jacks
- Left and right RCA audio jacks
- MIDI in/out (important if you want to record and play your own, or others', music using synthesizers, samplers, and so on)
- FireWire or USB
- Main hardware volume control (Most units are merely internal cards with PC-controlled software onscreen volume sliders only. A few offer external hubs with real knobs you can turn.)
- Second line volume control
- 1/8 inch line-in jack
- 1/8 inch microphone jack
- 1/4 inch headphone jack
- 1/4 inch line-in jack
- 1/4 inch second line in jack
- Dolby Digital (a surround sound technology)
- DTS ES (a surround sound technology)
- DVD-Audio (a high-quality audio spec; relatively rare software)

Chapter 7

- EAX 4.0
- EAX 3.0
- Remote control
- Specialized software (samples, such as E-Mu's famous library of rock, orchestral, and other sounds; E-Mu is owned by Creative).
- EAX
- "What U Hear" recording
- DSP
- Soundfont management

Savvy Tip

In addition to paying attention to the hardware features, you should spend some time listing what kind of software you'll need. Sound cards often come with quite a few useful utilities. When I installed my Audigy 2 ZS Platinum Pro unit, here's some of the software that showed up on my Start menu under Creative: Creative MediaSource (several utilities), AudioHQ, Diagnostics, Graphic Equalizer, DVD-Audio player, Minidisc center, speaker calibrator, speaker settings, surround sound mixer, wave studio, EAX console, and THX setup console. So take a look at what's bundled with the card you're thinking about buying.

8 CHOOSING AN ONLINE SERVICE

- ◆ Understanding playlists
- ◆ Apple's iTunes
- ◆ Napster's unique features
- ◆ Musicmatch
- ◆ MSN Music service
- ◆ Rhapsody
- ◆ eMusic for independents
- ◆ Comparing audio quality

Before turning from listening to music to actually making music in the next section of this book, this chapter explores the major legal online music services. I'm covering the main services here—those with large libraries of tracks from the major record labels and one special service, eMusic, that specializes in the independent labels.

USING MUSIC SERVICE PLAYLISTS

When you first install most music-listening software, such as the utilities that come with iPod or Windows Media Player, you're given the opportunity to have your hard drives searched and a database built of all your songs. This database generally includes information such as the artist, the song, and the album, along with additional details.

Subcategories, called *playlists,* can be built by you or automatically to group the songs into sets that you would listen to together. You can freely choose how you create playlists. You could build playlists called "Heavy Metal," "My Prom Music," "Recent Green Day," or whatever other subcategory interests you. This is similar to burning a custom CD for a party or for listening to while driving to work; it allows you to play DJ and choose the mix you like best.

iTunes for iPod

Other playlists can be built by the software itself, or online. Take a look at iTunes: www.apple.com/itunes/. Apple's iTunes music store (Mac and Windows versions) is the most famous of the online music stores. It was the first, it established the price point (99 cents a song), and it made downloading simple.

Click the Playlists tab and you see a variety of options. You can even share your special musical compilations by sending your "iMix" to the iTunes Web site. And others can rate your mix, letting you know just how good a DJ you are. Also take a look at the iMixes published by famous people like Minnie Driver, Juanes, and Elvis Costello.

Like most other systems, iPod can shuffle a mix automatically to freshen the play order. You can also automate playlist creation via the Smart Playlists feature. This is similar to those virtual taste algorithms found on Netflix, Amazon, and other consumer mega databases. You know: People who ordered *Bambi* also liked *The Lion King,* so maybe you would too!

A Smart Playlist is built after you specify a search filter, such as all the songs you've listened to in the past few weeks (iPod's musical odometer records the date and time you last listened to each song, and how often you've listened). Of course there are lots of other filters you can use: 1960s female country music and so on. Then, with your filter specified, iTunes searches far and wide and gives you a playlist you'll probably like.

Apple claims to have the largest online selection (1 million tracks) and permits you to listen to free, thirty-second previews of the music. The store says it has sold over 430 million songs, which is a pretty penny at 99 cents a pop.

I should mention a couple of additional qualities that the iPod system can boast about. So far, nothing *looks* cooler. And the fact that iPod and iTunes grew up together in the software labs at Apple means that they work very well together—making it very easy for rank beginners to manage their music. In fact, the iPod system is the most seamless, the best integrated, of all online music services—with the possible exception of MSN Music Service. Also iPod offers 8,000 audio books.

Understanding Napster's Radio

The reincarnated Napster is my favorite online music site for several reasons. You get to listen to streaming tracks or download (for later PC-based listening offline) unlimited tracks for no additional charge beyond the monthly $9.95 subscription.

It's slickly designed, allows you unlimited listening to a million *complete* songs (not the 30-seconds-only samples so common on other services). The big catalog of songs can be directly accessed (you can bring up most any song immediately by specifying track name, artist, or album). And it has some cool features such as building larger playlists around a few initial selections. Though it too has radio and "favorite playlist" features like others, you can freely listen to any song at any time.

Purchase the tracks for 99 cents if you want to own them. The Napster user interface is quite well thought out, and consequently it's very easy to search and browse. As Napster's advertising points out, you can connect Napster to your home stereo either via wire or Wi-Fi and use it like a huge CD changer. For an additional $5, you can freely fill and refill your MP3 player for no additional cost. To see about unlimited downloads, read the next section for details.

A Special Radio

Several music services offer variations on the playlist concept. Napster has playlists all right, and it also provides a "radio" feature that comes in handy, particularly if you sign up for the new Napster all-you-can-fit-on-your-media-player service.

The Napster radio has dozens of categories: the Velvet Lounge (sleazy listening), 15 Minutes (one-hit wonders), Radio Clash, and many more. You can regenerate the mix from a radio "station" anytime or save it to a playlist (click the button on the bottom of the radio list).

You can also create filters based on charts like *Billboard*'s, songs just added to Napster, artist, track, album, now streaming, a custom radio from your library, others' libraries, featured stations, top stations, Beastie Radio (they created a mix they think you might appreciate; I do), and so on.

Napster to Go: The iTunes Killer?

Napster has at least for a time knocked the wind out of the competition recently with its announcement of unlimited "refills" for $15.99 a month. No more 99-cents-each downloads.

Some critics have pointed out the curious math involved in legal downloads to portable audio devices. Consider this: the iPod 40 gig unit can hold about 10,000 songs. The hardware costs $399, but if you fill it with legally purchased downloads, the cost goes up by $.99 x 10,000 for a grand total of $9,900.

If you love your cool-looking iPod but like Napster's features, it is possible to indirectly get a Napster song (or indeed songs from other sources) onto your iPod. Burn a song to CD, then import the song back into iTunes to be able to move it over to the iPod unit.

Napster to Go claims compatibility with various MP3 players; to see the latest list, check www.napster.com/compatible_devices.

How does Napster to Go compare to actually *buying* songs? With Napster's system, you don't "own" the songs; you merely rent them and can keep them for as long as you pay the monthly fee. They "evaporate" if you stop paying the fee that feeds them blood to keep them alive. You can't burn them to CD or anything like that.

You can put the songs on up to three PCs and three portable digital units. One other issue: at this point only a few portable devices are compatible with the service (see www.napster.com/ntg.html for the latest list of compatible devices). The portable device must include Microsoft's Janus Digital Rights Management System. See if your unit offers a firmware update to include this technology.

Musicmatch

Musicmatch—a playlist-oriented, all-in-one digital music service has fans among critics in the press. I admire its effective user interface and the strength of its music collection, which is in my view second only to Napster's list of songs. If you want to try it, go to www.musicmatch.com.

Like many other music managers, Musicmatch assists you in organizing your tracks and also has features that allow you to burn to CD (or rip from CD), and of course play the music itself. Some people find Musicmatch's system more intuitive and seamless than other music management applications. Of particular value is the way that libraries (your personal library, recently purchased tracks, the entire Musicmatch catalog online) can seem to be a single, large library. For example, if you want to burn a CD made up of newly purchased and older MP3s, the process is painless compared to the approaches taken by some rival systems. But this online music service business has some pretty large companies battling it out for supremacy, so you get to watch capitalism on a rampage—but the results benefit us all in lower costs and greater quality. Put another way: by the time you read about the superior features or lower prices of one of the majors, you can bet that the others have matched the benefit, or are getting ready to.

Like some other services, there are free basic versions of Musicmatch programs and improved versions for some extra money. For example, the free version of Musicmatch Jukebox 10 gives you many features including access to the Music Store and On Demand (see below) and the capability of transferring music to portable devices. But get the plus version for a one-time fee of $19.99 and you add these features: custom library views, spreadsheet export, 48x burn speeds to CD (versus 8x for the free version) and 40x rip from audio CDs (versus 10x), a duplicate-track locator utility, special tech support, Radio Gold (see below) free for a year, Super Tagging (faster), and utilities that print covers and custom CD labels, as well as the ability to transfer your legacy cassette tapes and vinyl records to MP3.

In addition, there are other capabilities and optional upgrades, including

- *Musicmatch Radio.* Streaming music that plays through Musicmatch Jukebox. As you would expect, you can choose from among lots of "stations" featuring all kinds of genres. One version is free, but if you opt to upgrade to Musicmatch Radio Gold for $2.95 a month, you get CD-quality audio, no ads, remote access from any PC, and Artist Match, which permits you to request a station playing music only from a single artist or a group of similar artists.

- *Musicmatch on Demand.* For $9.95 a month (less for longer-signup plans), you get random access to any track of their more than eight hundred thousand songs. Using this service you can build a library of music that—unlike a radio station metaphor—plays precisely the songs you want, precisely when you want them. Included is the Musicmatch Music Discovery Engine, which matches your "patterns," your previous choices, against other Jukebox users. You get a list of up to one hundred of the artists most likely to please you. This is one way to discover new talent in those genres you most prefer. A similar utility, AutoDJ, has you type in a few of your pet musicians and searches both your library and the whole Musicmatch library to build a playlist likely to match your musical tastes. If you wish, you can further refine the playlist by specifying such elements as release-date cutoff, diversity, the size of the playlist, and the popularity of the songs themselves.

MSN Music

Microsoft's entry into any marketplace should never be ignored, and the MSN Music service is quite solid. They have some good exclusive content, intelligent categories, a large catalog, quality recordings, and—the Microsoft trump card—very smooth integration with Windows, Windows Media Center, and Windows Media Player. What else would you expect from songs encoded in Windows Media Audio? Actually, WMA, because of its superior compression—more tunes per megabyte, with no loss in relative quality—is used by most of the major music services (Musicmatch, Napster, and Rhapsody). iTunes of course uses its own proprietary Apple format.

Another aspect of MSN Music service's integration shows up in the seamless connection between their virtual radio and virtual store. You can be listening to a song and buy it with a couple of clicks. The process can be a little less direct in other services, but for most of us it's not all that difficult to achieve. After all, all services want to *make it very easy* for you to get your 99 cents from here (your wallet) to there. With Napster, for example, you just right-click a song in any radio playlist and choose Download Track(s). You can find the result in your library. You've got it. You can play it. To burn it, you'll be prompted to buy it. Happens in a flash. But never doubt Microsoft's ability to think up interesting twists. They offer special aggregations such as Atlanta in the 1980s, San Francisco in the 1960s, and surely Detroit in the 1950s.

You can sign up for commercial-free MSN Radio Plus ($4.99 a month) and pick from virtual stations, and so on. I found this approach a bit annoying because if you're interested in hearing a particular song, you have to listen until the station has it ready for play. True, you can click a button to move to the next song, but you cannot randomly access songs to listen to as you can with Napster (though you can randomly access songs to *purchase* at 99 cents).

Real's Rhapsody

Real offers 800,000 songs of what its Web site describes as "instant, digital quality." This makes me wonder what the writer of that blurb thinks that phrase is supposed to convey—what does *instant, digital quality* actually mean?

For those who enjoy the Real experience (I don't, though many critics in the popular press do), you can look it over at www.listen.com. I stay away from RealPlayer and other RealNetworks products at every available opportunity, but RealNetworks' offerings have an audience. Perhaps you'll like what they do.

eMusic: If You Like Independent Labels

These mainstream services are those that most people are interested in, but if you're fiercely avant-garde, like to be the first to discover the next Maroon 5 and introduce them to your crowd, or find independent spirits (pre-sellout) simply irresistible, try eMusic. They offer 500,000 (or perhaps 400,000—their Web site lists both figures) tracks from what they describe as "the world's top independent labels." It operates on a subscription system (40 downloads per month $9.99, 65 for $14.99, 90 for $19.99). You get MP3 tracks that you can burn as often as you want to CDs or transfer to any computer or portable media device, including iPod. If your tastes tend more toward a style of music rather than to specific artists, you might find that eMusic is a great value.

COMPARING AUDIO QUALITY

When you research the quality of the audio of the various services, you find three problems. First, they're in flux (they keep improving the specifications). Second, some services are deliberately obscure about this information, hiding it deep in an FAQ or elsewhere on their site. And third, there are two primary specs to consider in most cases: the download (or CD-burn) quality and the entirely separate stream (radio or song-sample) listening specifications.

Different online services' specs have been improving, but you might want to compare the other online services' recording specs to MSN Music's gold standard. It uses the efficient and respected WMA compression system and boasts a quality variable bit rate of 160Kbps, peak 256.

Napster, Real's Rhapsody, and Musicmatch (downloaded tracks at 160Kbps WMA files) are based on the WMA codec too. Rhapsody uses Windows Media version 8 (the current version employed in Windows is 10) and a proprietary streaming technology. This service burns CDs at 128Kbps. Listening to the streams from Rhapsody, those who have the Radio Plus subscription plan or All Access and who also have broadband get to hear both radio and music from their playlists at 128Kbps. Other customers hear playlists at that same rate, but radio is 20Kbps. iTunes and RealPlayer Music Store use the AAC codec.

Your pleasure at hearing the streaming audio from online sources of course depends first on the bandwidth of your Internet connection. But other factors do play a role. The iTunes protected AAC codec is considered by some to be of inferior quality to the WMA-based services. Real's Music Store is better than iTunes, according to some; it uses a 192 Kbps AAC-based codec. Napster and iTunes are said to offer lower-quality 128Kbps downloads. (Napster streaming was recently improved to 128Kbps.)

Whether these differences matter to you can depend quite a bit on your equipment, both personal (your hearing) and hardware (your playback devices). I suggest you listen to various streaming and downloaded songs and see if you notice a difference.

9 SOFTWARE THAT MAKES MUSIC

- ◆ When computers compose
- ◆ Music Minus One: Band-in-a-Box
- ◆ Algorithmic creativity
- ◆ ArtSong: where music meets graphics
- ◆ ArtWonk: you can spend months exploring
- ◆ Pirkle's classical approach
- ◆ Garritan Personal Orchestra: lush, convincing instruments on a budget

So far we've explored all kinds of ways to listen to music, essentially a passive, if often delightful, activity. But your *active* involvement in the music was limited to organizing, collecting, and otherwise managing other people's music.

Now the book turns to hands-on, active music making—improving other people's tracks or even composing and recording your own original music. In this chapter, you get your feet wet by exploring some interesting and useful music-generating software. You'll see how to seed, guide, or totally take control of the composition process.

In many of these intriguing music-generating applications, how much you contribute and how much the computer contributes is your decision.

But before you try your hand at composing and recording original music (or remixing existing songs into wholly new sounds) as you'll learn to do in upcoming chapters, why not see what the computer can come up with all by itself? Well, maybe with a little guidance from you from time to time if you wish to enter into the process.

THE COMPUTER COMPOSES

Can a computer write music? Sure, why not? Music is a combination of rules (such as orchestrations—choice of instruments—and chord progressions characteristic of, say, bluegrass) combined with chance inspiration. Computers are excellent at remembering and following rules (that's what computer programs are, just a list of steps to follow). In addition, computers can "roll dice," employing randomization to generate new musical "ideas." Blending the familiar with the new is a great recipe for music making. And with the arrival of MIDI—the technology that allows music to be deconstructed into its component elements—computer composition became not just possible but inevitable.

GIVE BAND-IN-A-BOX A TRY

To me one of the most interesting composition applications is Band-in-a-Box (PGMusic.com). You can download a Windows demo version from www.pgmusic.com/biabwindemo.htm (look in their Products location for the Mac version). After you download this demo, be prepared for a shock. This award-winning program is amazingly full of features you can explore, and the quality of its music can be a pleasant surprise. Band-in-a-Box (Biab) is to composition/accompaniment software what New York is to cities: in a class by itself, partly because you get so much band for your buck.

Biab can act as a quite competent backup band for a singer, guitar soloist, drummer, or whatever you do or play. Just mute the instrument *you're* playing, and click the play icon. They'll accompany you in lots of ways with often startlingly respectable chops. If you're a singer, mute the instrument that's carrying the melody if you wish, or sing along with it.

Which instruments are normally used in place of a singer? Sung melodies are the one difficulty for MIDI—even crudely imitating the human voice isn't yet possible, much less actual singing like kd lang. So MIDI songs usually assign the "vocal track" to instruments that are especially expressive, like the sax or cello.

Generate original music

But Biab is more than just a super karaoke system, or music-minus-one. It can generate all kinds of original melodies, harmonies, and orchestrations (there are loads of styles available, from Heavy Metal #1 to Chopin Waltz). Try feeding an Edgar Winter MIDI song through a Chopin waltz style, just for your edification. And if the built-in styles aren't enough, you can buy additional style sets from PGMusic, or even design your own custom styles. You can build personal styles from scratch using the Stylemaker, edit existing built-in styles (or combine elements of several styles into a new one), or even "auto-create" a style from an existing MIDI file. As you can see, Biab's capabilities are numerous, and the program offers multiple options for its many features.

This program has been around for many years—I remember using it when I got my first synthesizer back in the 1980s. And they've been constantly improving it for a couple of decades now. You'll likely be quite surprised at the depth and breadth of the features and impressed at the overall quality of the results you get. Whether you want to use it for music-minus-one accompaniment or for generating its own new melodies, embellishments, accompaniments, guitar riffs, improvisations, harmony (of various kinds), chord progressions, solos, and so on, it can be like sitting in with a group of extremely versatile musicians who are willing to try whatever musical ideas you come up with (or offer endless new ideas of their own). You can type in chord progressions (or melodies) of your own, import them from MIDI files, or let Biab automatically generate them (provide some guidance if you like, or use default styles).

Here's an example: I tend to like rock orchestrations, legato in the melody line, and songs that shift relatively often between major and minor chords. These are just preferences. Those and hundreds of other personal "rules" are easy to implement in Biab to guide the results of new melodic lines, solos, embellishments, and of course, to complete the song, the chords. To apply rules to melody generation (this is only one of many rule sets you can influence), choose Melodist from the Melody menu, and you see the dialog box shown in figure 9-1.

In figure 9-1 you can see many options, but I like to set the Change Instrument option to Every 8 Bars and the key to a minor key.

Now click the Edit button to further refine the chord and melody rules, as shown in figure 9-2.

In the editor shown in figure 9-2, I adjusted Unusual Chord Progressions from the default 0 to 30 and raised Mix Major and Minor Progressions from 0 to 40, just because I like the kind of music these rules generate.

Click OK twice to close both dialog boxes, and at once your new song starts playing away. If it sounds like a bad night at the local Holiday Inn's Lizard Lounge, just click the Style button shown in figure 9-3 and try some different orchestrations.

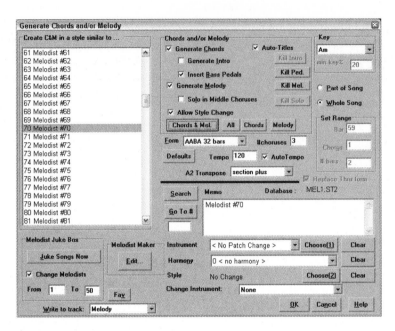

Fig. 9-1: Here's Band-in-a-Box's main melody- and chord-generating dialog box.

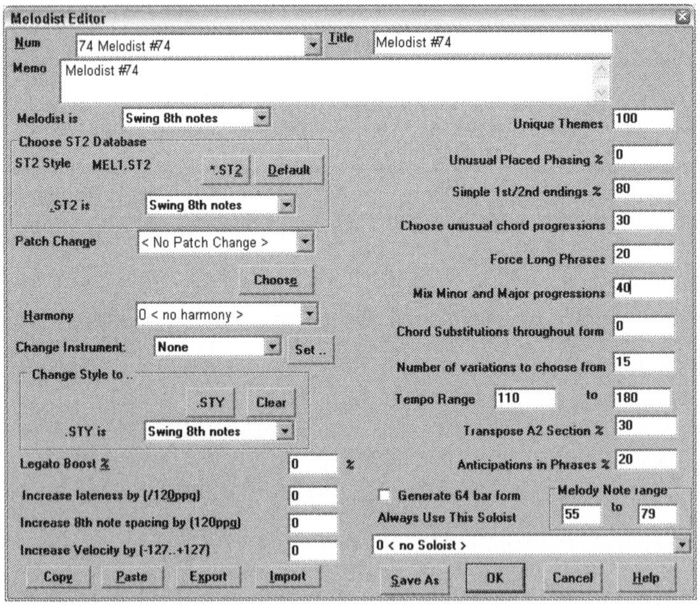

Fig. 9-2: You can specify additional parameters for your new, auto-generated composition in this Melodist Editor in Band-in-a-Box.

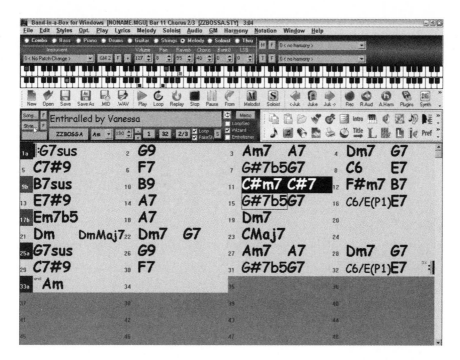

Fig. 9-3: This is the main Band-in-a-Box editing window, where you can adjust styles, key, orchestration, FX, and many other aspects of the current song.

What's Most Cool?

What's most cool about Biab? It's loaded with so many features and options that you can lose yourself in it for months. And unlike sequencers where you sometimes must start from scratch, the virtual "musicians" in Biab are always ready to supply endless accompaniments and inspiration. You contribute as much or as little as you want. It's a musical wonderland in many ways—a playground where you can get accustomed to some of the great things you can do with computer music. Import high-quality MIDI arrangements that someone else has already created, if you wish, from the various MIDI sites mentioned in this book (http://adventurelandtravel.com/FreeMidiMusic2.html has some brilliant ones, for example). Then just play them in Biab, or save them to .WAV files or .MP3 to burn to CD's or add to your portable MP3 player's library. Or fiddle around with new Biab "styles" applied to downloaded MIDI. Or record or type in your own music. Whatever path you take, you're likely to stay up late having great fun.

Here's another use for Biab: If you ever need some original music for a local TV ad, a custom DVD, or some other purpose, you can purchase a track from a canned-music library (any TV studio has these discs). But that's relatively expensive, and the music is often less satisfying than what Biab can generate for you in seconds.

Try another experiment. Import the chords from a familiar MIDI song then have Biab generate a brand-new melody for those chords. Choose Import Chords from MIDI File in the File menu. Click the Open button in the dialog box.

Now that you've got the song's chords (and usually the melody as well) loaded in Biab, you can play around with the orchestration by clicking on the individual categories (such as Guitar or Strings), then changing the Instrument in the dropdown list in the top left of the screen. Or make wholesale changes to the orchestration and rhythm by changing the Style (use the Style menu or the Style button shortcut).

Try generating an entirely new melody now. Stop the song from playing by clicking the black square icon. Click the Melodist button. Click the Melody button on the dialog box to deselect Generate Chords—you want to use the familiar chords you imported from the song. Click the Edit button if you want to further specify how the melody should be generated. When you're satisfied, click the OK button. The melody is generated and now plays against the original chords. If yours was a wonderful song choice, perhaps the melody generated by Biab won't equal the quality of the original, but you can always try generating another one. Just click the Melodist button, then click OK to close the dialog box. A new melody is created that fast.

Biab ships with a generous set of "soloists." Evidently the program's creators sit around and closely analyze the elements of a famous musician's technique—that of a great pianist, for example. In a separate utility from the Melodist you can specify a "soloist."

Biab isn't a sequencer, so it doesn't specialize in some aspects of music recording, such as employing audio tracks as opposed to MIDI (see chapter 10 for more on sequencing). You can, though, record what you play along with the "band." But I've only been able to give you a broad outline of Biab's capabilities. Do download the demo and spend some time exploring it. If you have any interest in orchestration, computer composition, improvisation, or being the "one" playing with a music-minus-one group, prize-winning Biab may well be the software you've been waiting for. You can find it at CompUSA, so it's pretty popular.

ARTSONG CROSSES THE BORDER

Digitization has allowed musicians to manipulate even the tiniest, subtlest aspects of music. Once quantified into a MIDI file, everything from vibrato to global "humanization" (introducing deliberate perturbations to the dynamics, rhythm, or other aspects of a track) can easily be applied. Music, in MIDI, is expressed *mathematically* and can thus be manipulated quite efficiently by computers.

WHERE TO GET GREAT MIDI SONGS

You can easily find zillions of MIDI files, many quite excellent, by searching the Internet. But if you just type MIDI into Google, unfortunately you get mostly junk sites. So instead, try using my favorite MIDI file search engine: www.vanbasco.com/midisearch.html. Type in a song's name, or an artist's, and you'll get links to often excellent MIDI arrangements. Some of them are better than the original. You'll usually get several different versions of a particular song when you use this search engine. If you have time, listen to them all, but if you want to download the best version, your best bet is to choose the largest file. If you find one version that's 12K and another of the same song that's 97K, well, the person who created the larger file probably did a better job. This isn't always true, but often enough.

Another tip: If you look at the piano roll view, or track view of a MIDI song file in a sequencer, you can sometimes see that it has complicated, subtle changes in tempo, volume, pitch bending, or other nuances. This usually means that the author of the file took pains to add realism to the various instruments and perhaps "humanized" the rhythms. In other words, the tracks aren't perfect, machine-like, robotic sounds; instead, time was taken to make them more musical. And of course you'll certainly hear the difference when somebody created a great-sounding MIDI song. You'll be surprised—there are thousands of commercial quality, first-rate MIDI songs out there waiting for you to find them. And when you find a couple of good ones at a particular Web site, add it to your Favorites bookmarks in your browser. After a while, you'll have a collection of perhaps 50 excellent sources of great music.

Something similar happened when digitization led to the computerization of graphics. It's easy to adjust all the elements of a photo, down to the smallest quantum, the pixel. Doubtless you've boosted the contrast in a muddy photo or adjusted the tint. But did you consider that the algorithms that transform graphics are similar, sometimes identical, to the algorithms that transform music? What's more, you can cross the border between graphics and music in various ways (you've seen the "visualizations" that display a dynamic graphic that changes in response to the music being played in Windows Media Player and other applications).

What would happen if a music program analyzed qualities in a picture and used that visual data as the basis for creating music? That's one of the interesting features of ArtSong, algorithmic composer software that, in its own way, offers loads of flexibility for those of us interested in computer composition. Among other things, it explores synergies between music and graphics. (See figure 9-4.)

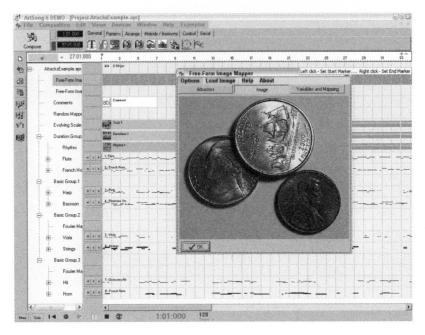

Fig. 9-4: ArtSong can turn pictures into music.

Feed it some pictures, and listen to the varying results. (It generates music in many other ways, but let's fool around with this interesting technique for a minute.) Download the demo version from www.artsong.org/download.htm. Then try the Image Mapper feature. Also try experimenting with some of the songs on the Examples menu. AttacksExample is a particularly interesting orchestral invention, and the SimpleCannon illustrates how well this program can generate effective polyphonic melodies.

ArtSong, like the other applications discussed in this chapter, benefits from increased exposure. The more you noodle around with it, the better you understand the effects of the pattern generators, arrangement features, textures, and so on. The program's documentation points out that "ArtSong is 'intended' to be inherently devoid of any particular musical style; an electronic-procedural manuscript paper that you can apply to any specific composition problem. Your creativity, selection of, and skill using the various components, algorithms, composition-variables, control-events, etc., determine ArtSong's musical output."

Indeed, what you put in determines what you get out. It's quite versatile. You can feed it an existing Bach fugue in a MIDI file, graphics, and many other starting "seeds"—there's even a feature allowing you to write Basic or Pascal scripts to manipulate the music. Templates and presets also provide you with a way to save favorite setups.

This, like Biab, is a program offering lots to explore and rewarding that exploration with compositional tools you might use in the future. Computer composition can provide you with many new musical ideas—and ArtSong is no exception. I suggest you start your exploration of ArtSong by opening the Help menu and locating the Tutorials in the Help system. These lessons open the doors to the program's powerful feature set.

COMPOSE WITH ARTWONK

ArtWonk is another computer-based composition system that can employ art—and lots of other things, including stock market data, even DNA—as inspiration for its creations. Computer compositions have to start somewhere, even if it's a random number to seed an arpeggiator. But anything from a Beethoven piano concerto to your sister's e-mail can be used by some of the programs in this chapter.

ArtWonk specializes in a kind of advanced step-synthesis "device" that you design from an extensive, interesting toolbox. You can select from clocks, arrays, widgets, custom scripts, sequencers, logic units, patterns, and graphics (hence the name ArtWonk). You can also go the other way and generate graphic images or video based on the music or, as with most aspects of ArtWonk, in hundreds of other ways as well. This software permits microtunings, custom functions and macros, and many other widgets, gadgets, and virtual tools that make putting together a new composition—music or visual—a real pleasure. It's well thought out and, once you get the hang of it, efficient. The excellent online Help system describes the overall behavior of ArtWonk like this:

> You compose by connecting "modules"—graphical objects that represent functions and processes—in real time, adjusting the parameters as you go. ArtWonk will drive MIDI synthesizers, soft synths and sound cards directly, effectively "playing" them in real time or optionally responding to user or environmental parameters; it will also record directly to a standard MIDI file. Real time paint graphics can be created and manipulated on the fly; and both music and paint can be created together as a synesthetic whole.

As you can see, their bold (and well-realized) goal is to cross the graphic/music frontier in powerful ways. Download a fully functional version for a thirty-day trial from http://algoart.com/download.htm.

The primary ArtWonk screen is divided into zones representing the process you follow from idea to music. A light blue area—the largest area by default, though you can rearrange the screen—is where you build your "patch" by right-click adding, moving, and connecting various modules. At the top is a dark blue panel where you can build your controller by adding buttons, sliders, a mouse area,

LEDs, text, meters, knobs, and so on. It's a bit like building a hardware unit—but a lot faster. And if you try to make an impossible connection, it's quietly refused rather than starting to smolder like true hardware might. Finally, on the left side is a tree view providing another way of looking at your patch's modules. There's also an optional graphics window where lovely results can appear—particularly if the input is symmetrical, fractal, or otherwise destined to please the eye.

ArtWonk was designed to allow you tremendous freedom and to respond to many "rules" or input. You could, with this software, play a MIDI keyboard or other MIDI source, while ArtWonk follows along harmonizing with your inventive melodic lines or displaying visual forms and colors representing the music as it flows along.

This software actually approaches fabulous, but you do have to settle yourself down and get to know it—you can't immediately produce masterpieces of cross-disciplinary musical graphics, or graphical music. The program is deep and rich, and you have to experiment with it for a time, and learn a thing or two, before it yields treasures reliably. For an example, see figure 9-5.

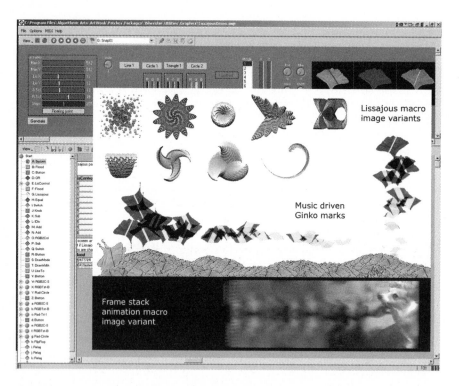

Fig. 9-5: These lovely lissajous images are all variants derived from a single macro using different settings. Courtesy of Professor Jamy Sheridan, chair of the Department of Experimental Animation at the Maryland Institute College of Art.

TRY PIRKLE'S CLASSICAL APPROACH

Perhaps you can tell that I'm a fan of interesting computer-music-generating software. Let's look at one more good one before this chapter concludes: Pirkle's Music Composition Studio (PMCS). It's sure to be popular with people who enjoy classical music, though the system can also generate other musical genres from blues to reggae. To get an earful of what it can do, listen to the music samples at www.pirkle-websites.com/composer.

The user-interface (see figure 9-6) takes a little getting used to because it's not the Windows-style screen full of buttons, sliders, text fields, and other input controls that we're all used to. Instead, it steps you through a series of questions (and at each step you provide your answers). This is going to be a familiar, linear data-entry process to those who remember life pre-Windows, when DOS was king.

After you spend a few minutes answering the questions, the program goes into machine-intelligence mode and builds an often remarkably listenable piece of music. It's original, but you can tell it was intelligently composed, if by a machine. And, given that you participated in the decisions that led up to the result, why, you're a bit of a co-composer along with your computer.

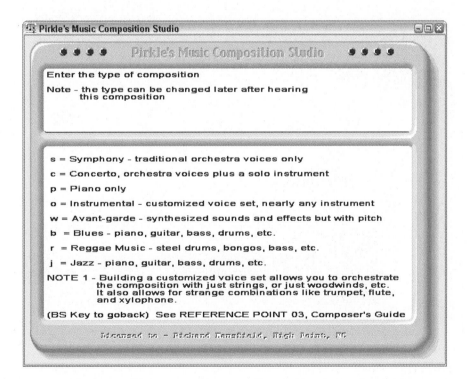

Fig. 9-6: Pirkle's Music Composition Studio's deceptively simple user interface conceals a sophisticated music-generating application.

SAMPLE GARRITAN PERSONAL ORCHESTRA

I enjoy pairing Mr. Pirkle's excellent classical-music generator with the equally impressive set of orchestral samples available in the Garritan Personal Orchestra, shown in figure 9-7.

Garritan's very high-quality—and, at $250, very low-cost, compared to the competition—four-CD set of samples even improves the sonics of rock songs and other genres. You get the full orchestra, from convincing timpani to lush cellos, as well as ensembles such as string sections and quartets. You can, of course, also layer your own custom ensembles, though I suggest you beef up your RAM memory to 1 GB. You can add expression to the instruments in various ways, further increasing the realism of these sounds. Have a listen to some demos at www.garritan.com/mp3.html.

In addition to all the expected instruments, there are some special gems, including a terrific Steinway grand (several versions of solo and duo), a beautiful solo Stradivarius, and a realistic, bold pipe organ.

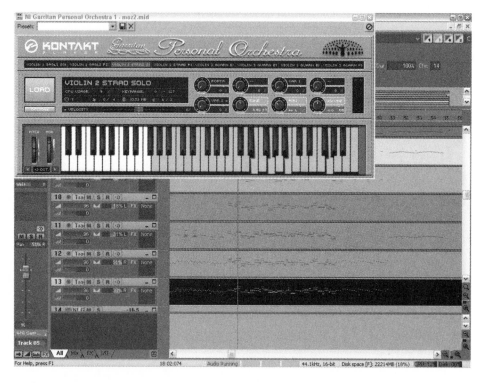

Fig. 9-7: Now you're a threat to Mozart himself . . . well, maybe to Salieri anyway. Whatever music you play through the Garritan Personal Orchestra sample library sounds great.

The key point is, almost every sample in this library is extremely well done, and the samples add quite a bit to your musical efforts. No more cyclic ringing from too-short samples of cheesy string pads, no more brass that sounds too brassy, or other artifacts and defects. Add a Garritan French horn to your song and it sounds *just like a French horn*, not some synthetic, honking caricature.

When you get this set of samples, you get an orchestra at your beck and call, something people have dreamed about for centuries. You can demand, as only kings once could, that a string quartet play your new composition. You can even summon a full orchestra with the snap of your imperial fingers, then complain that there are "too many notes" in this particular concerto. What fun you'll have with a personal orchestra.

10 SEQUENCING

◆ Understanding sequencers

◆ A music studio in a box

◆ What MIDI does

◆ Creating a finished song with a multitrack sequencer

◆ Manipulating tracks

◆ Humanizing

◆ Orchestration

◆ The mixdown

◆ Understanding effects

◆ Sequencers, good, better, and best: PowerTracks, Cubase, Sonar 4, Pro Tools

For many readers, this chapter will be the heart of the book, the starting point for making your own music. Instead of merely playing DJ or putting together playlists and custom CD compilations of other people's music, you can get in there and *manipulate the music itself.* Or even compose, play, and record it. It costs surprisingly little these days to get started with a virtual recording studio and a massive collection of virtual instruments that sound real, or unreal, as the spirit moves you.

In this chapter you see how to download a great song in MIDI format, then, using a powerful free demo sequencer, change the instruments used in the song, add reverb, humanize it by manipulating the rhythm, doubling the melody, adjusting the melody's pitch, or changing the dynamics (loudness) of some notes randomly, and finish it off by recording it as an audio .wav file, adding reverb, a gate, an expander, compression, and EQ like the pros use. You're likely to be startled at how much you can do and how quickly you can do it with the computer's assistance and with the powerful software available today. When you're done, the song will have been transformed into something closer to your taste in music, and you'll have been the arranger, producer, and mixdown specialist, among other roles. But your primary achievement will have been to move from passive listener to active participant, and you'll be directly responsible for the quality of the final song.

After learning these techniques, you can even load in your favorite music from CDs and adjust elements of the sound by adding effects like additional reverb, flange, wah-wah, vocoder, exciter, or any other effect you like. If you wish, you can record yourself playing an instrument along with the MIDI tracks or singing along. The ultimate in hands-on music, of course, eliminates the downloaded MIDI song and starts from scratch—recording you or your band, mixing, adding effects, and turning out a potential hit song.

All these levels of participation are available; the results depend on how much time you have to learn to use these musical tools and, of course, on your talent. If you do come up with something great, there are ways to get it out to the public, and maybe even a record company scout—put it on the Internet and see what happens. Visit www.indiemusic.com to see some resources you can exploit, or go to PureVolume.com and CDBaby.com.

But we're getting ahead of ourselves. Presumably you need to begin with sequencers, samplers (see chapter 11), and some other technologies before you're a direct threat to Sheryl Crow.

What Sequencers Do

Sequencers—or, if you prefer, *multitrack software*, since most of today's sequencers come bundled with everything but the kitchen sink—allow the musician to record, mix, edit, and otherwise manage a musical piece. They're also sometimes called digital audio workstations (DAWs). But I prefer to stick with the classic term *sequencer*, as long as you bear in mind that lots of tools, extra features, and sometimes freebies like sample libraries and effects plug-ins now usually come bundled with sequencer software.

Today's sequencer is to a song what a word processor is to a document: a big set of specialized tools to help you shape your masterpiece, near masterpiece, or total disaster, as the case may be.

Sequencers often include lots of special features such as notation facilities, step recording (so you don't have to play in real time if that's a challenge), event list editors, plug-ins and effects (like compression, chorus, echo, reverb, and so on for individual tracks, or to the final mixdown), special mixing features such as automated faders, various "views" of the song (such as piano roll, MIDI event list, traditional sheet music [staff], loop editor, audio waveform editor, mixing console, synth rack, and so on).

GETTING STARTED

If you've not used sequencers before, I suggest you buy the inexpensive $49 PowerTracks powerhouse described below; it includes all the features you'll likely want to experiment with. If you do end up wanting other capabilities, you can always replace it with a more official, more expensive product like the excellent Sonar 4 from Cakewalk.

Hardware sequencers exist but are now far less popular than the more flexible and generally less expensive software sequencers. What used to be a stage full of keyboards and effects boxes has now all moved into the computer. Sure, there's still a keyboard in the band, but often only *one*. All the rest of the synths and effects hardware have migrated into "virtual" machines inside the PC. Your computer can become a desktop studio, capable of producing professional-quality CDs. The day of the hissy, muddy, low-quality cassette demo is over. Also, some hardware synthesizer keyboards have sequencers built in, and more advanced keyboards (usually called *workstations*) always include a sequencer.

Some sequencers do highly specific jobs, such as drum machines or loop-based editors. We'll explore sampling and loop-based applications like Acid and Live—a hot area of music today for those into electronica, hip-hop, and so on—in the next chapter. Many of those types of sequencers are designed to be actually "played" in real time, as if they were instruments themselves. In this chapter we'll focus on general-purpose, conventional software sequencing, your primary music manipulation toolbox for more traditional musical genres, such as jazz, classical, or rock and roll. Yes, geezers, rock is now a traditional music form, a period in music history.

The better sequencers permit you to record and manage actual audio tracks (such as .wav files) right along with MIDI tracks. Others are limited to strictly MIDI tracks, so be sure you get what you need before spending money (and time learning to use it) on software that won't ultimately do what you want. Other sequencers, such as Pro Tools, are popular among professionals but perhaps a bit high-end for home, or even some commercial, use. Other sequencers go beyond music sequencing to include suites of utilities for video postproduction, scoring, and so on.

PowerTracks Free Demo for Beginners

If you have little experience with sequencers, I suggest you get your feet wet with PGMusic's PowerTracks sequencer. It offers a large number of features, is friendly to beginners yet packed with powerful tools, and you're not going to break the bank with its $49 price tag. To try the experiments in this chapter, you don't even have to spring for the $49. You can just download the free demo and follow the step-through that takes you from MIDI song through the mixdown and mastering process to final audio .wav file ready to burn to CD.

True, you can download various shareware and freeware sequencers, but they can be disappointing—particularly because a truly useful sequencer is a complicated set of tools, and you need support while climbing the learning curve. PowerTracks has lots of support behind it, including an excellent manual and a very active, friendly, and knowledgeable online forum where an expert will often answer your question within hours after you post it.

PowerTracks is a seasoned product and an exceptional value. It's produced by the same people who sell Band-in-a-Box, described in the previous chapter. I have no connection with this company, other than admiration for the powerful, well-supported software they sell at an incredible bargain. To my knowledge there's nothing else like it to help the beginner get started with a home music studio.

Understanding Sequencer Features

A while back a sequencer was rather simple: you created or loaded a MIDI file and then played it back through a mixer. Now, though, sequencers are your central music-generating engine. True, you still may need an actual keyboard or other musical input device and a few other pieces of hardware like speakers. But when you fire up a general-purpose sequencer such as PowerTracks, Sonar 4, or Cubase, you're looking at a combination of tools that were previously separate hardware units (such as mixers and multitrack tape recorders) and software utilities (such as audio wave editors or effects), which are now merged into the sequencer package.

A decade or two ago, computers just weren't powerful enough (neither speedy enough nor with enough memory) to host a virtual music studio. If you wanted to set up a music studio, you had to buy lots of hardware costing thousands and hook it all together with *wires* in *racks*:

- Effects units such as reverb, echo, limiters, and others
- High-quality professional instrument samples in ROM banks, such as the famous Proteus rack hardware units from E-Mu

- Samplers to record your own samples (either an individual sound such as a cowbell, or a copy of a bit of music that can be looped [see chapter 11])
- Synthesizers filled with artificially generated instrument sounds, some that mimicked acoustic instruments and others that were purely artificial like the theremin or Moog sounds
- Drum machines
- A multitrack tape recorder
- A multitrack hardware mixer to attach to the tape recorder for multitrack recording, or to blend the tracks down into a final stereo mix while routing and applying effects
- A computer to control *some* of these devices via relatively simple software sequencers

But over the past 20 years, computers have gotten far more powerful, and what used to be primarily hardware peripherals have migrated into the computer as *software*. All these components are now commonly collected into a general-purpose sequencer application.

Most software sequencers today can also manipulate and embrace plug-ins and other component software—even from other manufacturers—that's connected to the primary sequencer. Components include sound banks, "soft synths," harmonizers, and other effects via DXi, VST (Virtual Studio Technology, created by Steinberg, which generates sounds in real time), and other technologies and standards that help tie everything together under the governance of the primary sequencer application and its suite of utilities.

What Is a Sequence?

A sequence is essentially a *track*—usually a recording of a single instrument. For example, if you have a keyboard synthesizer with MIDI out, you can select the trumpet voice and play taps while the software sequencer listens and records your performance. MIDI records all kinds of information about that performance (it's not audio; when you play back the MIDI track, the data is sent back to the synthesizer where audio is generated based on that incoming data stream). Though your synth may or may not be capable of detecting subtleties such as pressure or aftertouch, MIDI can record that kind of data if your synth does transmit it.

After a track is recorded in MIDI, it can be played back in the sequencer and a MIDI instrument (or more likely a software sample in ROM or a soft synth) reproduces what you originally recorded. During this playback, you can be recording a different track simultaneously, thereby building an entire set of tracks that can be played back as a complete, multi-instrument song. Or, of course, you can go online and download from hundreds of thousands of MIDI (.mid) files that others have recorded. Many are excellent, and some—well, not too much effort or talent went into them, shall we say. Search for online MIDI music via www.vanbasco.com/midisearch.html. You can also find

Savvy Tip

Companies such as M-audio are now making small, two-octave battery- or USB-powered keyboard MIDI controllers. These are lightweight, fit into a laptop bag, and have knobs, expression wheels, and footswitch ports for real-time control of many MIDI parameters. Units like M-Audio's O2 or Ozone are quite handy if you want to edit, manipulate, or add new bass lines or other tracks to MIDI songs. And of course you can use them to add expression to existing tracks via pitch bend, vibrato, aftertouch, and any other MIDI-controlled musical quality.

first-rate professional quality MIDI recordings, for a price ($3 per song in packs of ten). Listen to some of the demos at www.trantracks.com to see what I mean. This site has the most consistently high-quality MIDI songs I've found. Another site offering high-quality MIDI songs is www.cyber midi.com. And it's quite reasonable ($44.99 for a year's unlimited downloads).

In addition, some sequencers permit you to record *live* instruments—not via MIDI, but through a microphone. These are recorded usually as .wav files. These .wav tracks can, in better sequencers (including the $49 PowerTracks), reside right in the sequencer alongside MIDI tracks. This way you can blend live with MIDI tracks into your final mix.

When you've finished and are satisfied that you've quantized, transposed, randomized (the opposite of quantized; randomizing throws the rhythm or dynamics a little bit off, to "humanize" the sound), added effects, and whatever else you want to do to massage the tracks, you then use the mixer feature in the synthesizer. This mixer allows you to add additional effects including global effects to the entire orchestration, such as reverb, and mix down all the tracks into stereo (or, increasingly, into surround sound). Then you can add finishing touches (see "Mastering" below), burn the result to CD, and mail it off to Sony Music so they can start the bidding war to sign you.

MIDI Data Streams

MIDI data sent when you press a key on your synth can typically include:

- Patch: the instrument
- Note duration
- Velocity: how fast, thus how hard you hit the key, resulting in a louder, or sometimes brighter, note
- Aftertouch: The more sophisticated keyboards detect what your finger does following the attack, or the initial hit on the key. Experienced musicians can use their fingers to manipulate the sound, adding realism, in various ways depending on which instrument they're play-

Savvy Tip

One of the best ways I know to humanize MIDI songs is to apply some of the presets available in Jasmine Music Technology's Style Enhancer software. This little gem really does what it promises. It takes mechanical-sounding bass lines, or piano (or whatever), and adjusts tempo, aftertouch, velocity, pitch bend, and other qualities to make the instruments sound nuanced as if they were really played by a human. You can create your own presets, or use the many included presets such as Solo Lyrical Bending (for guitar), Slap Intensive with Special Gliding (for bass), or Jazz Rock Drums (for Steely Dan). The Style Enhancer also includes other utilities, such as time morphing and autophrasing. But you should really give their modeler your initial attention. It can greatly improve stiff, robotic MIDI (which is all too often a problem with MIDI files you download). Click the Midi Process menu, then choose Modeler and click the Styles dropdown list or the individual dropdown list below it. Try out the Style Enhancer by downloading the demo from www.indiancanvas.com/se40_demo.htm. Another really useful tool for MIDI work is Jasmine's Onyx Arranger. It includes harmonizing, humanization, and auto-arrangement features you'll want to explore. Check out the video demo, then download the demo software, from www.indiancanvas.com/onyx.htm.

ing. A violinist, for instance, often moves his fingers to cause the sound to waver in pitch, or a guitarist slightly bends the note. Aftertouch can usually be programmed to produce filtering, pitch, vibrato, or loudness changes.

- Pressure: an aspect of how you press a key that's similar to aftertouch. Sophisticated keyboards detect how your finger presses the key *while* the note is being played, and produce vibrato, tremolo, and added brightness or volume.
- Pedal: legato, sustaining notes while you depress the pedal, as on a piano
- Pitch bending: usually controlled by a wheel on the keyboard
- Modulation: usually controlled by a wheel on the keyboard. It can be set to control various effects such as adding tremolo, vibrato, pressure, and so on.

A sequencer holds multiple tracks, often one track for each instrument played in the song (though a single patch can be a sample of an ensemble of instruments, like string quartet, brass section, or orchestra hit). And recall that many sequencers allow you to also add "audio" tracks; waveforms are usually visible in the track rather than dashed lines, piano rolls, or other ways of indicating MIDI data. So you can mix down to stereo audio, add loops and samples (from, for example, .wav files), import or record in real time additional MIDI tracks, and so on. The flexibility can be quite overwhelming at first, and quite welcome down the line once you've got a feel for your sequencer and its ways.

GETTING A FEEL

Let's do some sequencing, step by step, so you can get an idea how to produce (*generate* might be a more accurate word) some really good-sounding music. After you've had more experience, you'll earn the title *would-be producer*.

Start on the Internet at this location to find a really good .mid file: www.vanbasco.com/midisearch .html. The quickest way to produce a song is to find a nice MIDI file that has already been recorded. Your job then isn't to be the musician but rather the arranger and generator (or producer). You'll decide the relative volume of the tracks (in the mixdown phase), correct flaws like bad rhythm, perhaps choose different instruments (orchestration), and fiddle around until you get the sound you like.

Given that everyone's taste is different (and some people seem to have more taste than others), what you do with a song might be way too electronica for me—too mechanical, reductive, and repetitive for my tastes. By contrast, my mixdown of the same song might seem to you way too acoustic, orchestral (classical sounding because a violin section pad got in there), too "loose" (some notes will actually be off the beat!), and generally not robotic enough to induce a trance at a rave. Whatever. That's the nice thing about taking control of your music. With a good MIDI file, you've got a great start, but ultimately you take the music in your personal direction and end up with a custom remix that gets your friends so worked up they cry their heads off. Whatever.

The Internet is full of original (as well as famous) songs recorded into MIDI files. Many of these files are of exceptional quality, recorded by real musicians, and even well orchestrated. Now and then you find a surpassingly beautiful MIDI file (Tran Tracks's, work, mentioned earlier in this chapter, is especially reliable). You can also find some surprise treats. One MIDI version of *Losing My Religion* is reportedly sequenced by none other than Mr. Stipe himself.

We'll do a little orchestration and a few other tricks during our step-through. So, download five or six .mid files and double-click them to listen to them in Windows Media Player or your favorite media software. When you find one you like, load it into your sequencer and we'll get started. For this example, I'll use PowerTracks because it's a very inexpensive way to get your hands on lots of sequencer, fast. Download the demo at www.pgmusic.com/ptdemo.htm. The demo lasts quite a long time, though the Save and accessory plug-ins are disabled. After downloading, you'll be able to follow along with this example.

The following steps give you a general idea of how to transform raw tracks into a finished, polished, final mix you can burn to CD. In the section titled "Special Refinements" following these steps, you'll find more details about some of the steps in the process.

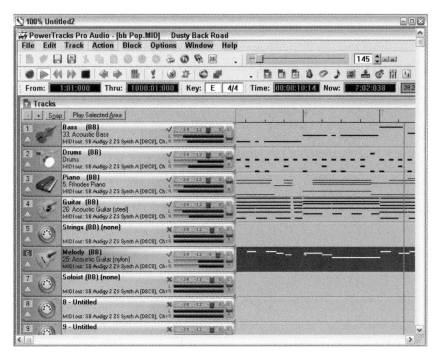

Fig. 10-1: In this PowerTracks screen there's lots of information about the music, including icons to help you immediately identify the instrument playing on each track.

OK. You've got your file loaded and you can view the MIDI tracks, as shown in figure 10-1.

1. *Listen.* The first step is to listen to the song. See what you don't like about it. Right off the bat I felt that the melody line, an acoustic guitar, was an octave too high. This is not a surprise—many MIDI files you download contain tracks that are too high. The melody track is often an octave higher than you'd like. I don't know why this is. Even some other tracks such as the bass track can require transposition, though less often. Another strange, but often predictable, adjustment that nearly always must be made: the drum track is too loud, so you have to reduce its volume. And some people put lots of reverb on every track, but I find it makes some instruments too muddy, particularly bass guitar.

2. *Transpose.* To lower the melody, while keeping it harmonic with the rest of the tracks, you can transpose it -12 steps. Click the track to highlight it, as shown in figure 10-1. Click the << two green back arrows on the toolbar at the top to move the selection point to the start of the song. Choose Edit, Pitch Transpose. A small dialog box opens. In the Value field, type -12. Click the OK button. Press the space bar to listen and you'll hear the transposition in the guitar. If you don't like the effect, Ctrl+Z undoes the transposition.

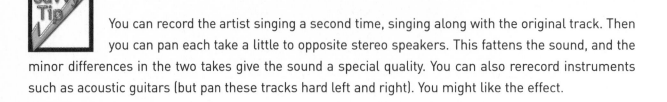

You can record the artist singing a second time, singing along with the original track. Then you can pan each take a little to opposite stereo speakers. This fattens the sound, and the minor differences in the two takes give the sound a special quality. You can also rerecord instruments such as acoustic guitars (but pan these tracks hard left and right). You might like the effect.

3. *Double.* Often doubling the melodic line adds depth and interest, a nicer feel, as long as the copy is slightly different from the original. With the melody track highlighted (selected), click the << reset-to-start icon on the toolbar. Press Ctrl+C to copy the entire track. Now click an empty track (number 8 will do in my example), and when that track turns red (showing it's selected), press Ctrl+V to paste the copy. With your new track still selected, you can transpose to add harmony or, as I like to do, subtly adjust the velocities and rhythms of the second track to make them just perceptibly more human.

4. *Humanize.* Choose Edit, Randomize. The dialog box shown in figure 10-2 appears. You can try (and if necessary undo) various settings, but I've found that the offsets (before and after the beat), and the adjustments in attack (velocity) and duration shown in figure 10-2 work pretty well.

Now your copied track is a little different from the original, and if you look closely you can see subtle shifts in the beat and duration of the notes, as shown in figure 10-3.

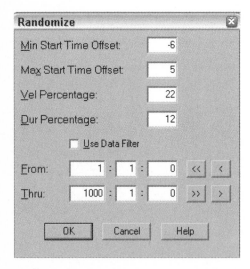

Fig. 10-2: Here's where you can slightly distort a MIDI track, to give it a more human (less beatbox) feel.

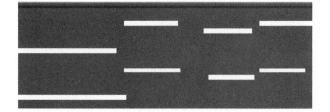

Fig. 10-3: After humanizing (randomizing), these two strings of MIDI notes are slightly off beat from each other and have slightly varying durations. Look closely and you'll see the effect; listen closely and you'll hear it.

Savvy Tip

Using randomization or humanization features pleases some people, but if you're into electronica or other contemporary music where a robotic feel is wanted, you can go in the other direction and eliminate variations by applying the quantization or quantize feature (also found in the Edit menu in PowerTracks).

---

5. *Orchestrate.* I decided that the default Acoustic Bass didn't best serve this song, and it's easy enough to change the instrument. (See figure 10-4.) Right-click the name Acoustic Bass in the track, then move your mouse pointer to the Program entry and continue following the pop-out menus until you select a different instrument. In my case, I chose the fingered electric bass. You can listen to the instrument playing all by itself by clicking the blue S symbol (for solo) on the toolbar at the top of the screen. Then whatever track is selected (red) plays alone. You can also solo additional instruments at the same time, so you're hearing only those selected tracks.

After you've chosen a different instrument, a new icon appears in the track to clue you visually, as shown in figure 10-5.

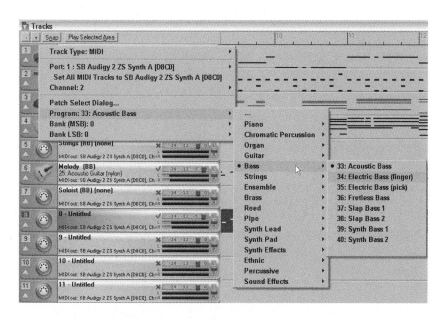

Fig. 10-4: Change orchestration with these pop-out menus.

Fig. 10-5: This is the symbol in PowerTracks for the fingered bass.

6. *Mix down.* Play back your song and listen to the relative volume of each instrument. You now create a stereo mix from all the other tracks. You can adjust the pan (left or right position) of each instrument track, as well as adjusting its volume to keep its sound in a pleasing relation to the other tracks—usually not overwhelming, and not lost too far down either. Generally speaking, the melody is the loudest and also resides right in the middle of the stereo spread.

7. *Master.* After you've got your stereo (or surround) final mix, you should take one final step. It's called *mastering* and it can mean adding some special effects to the final version before it's burned to CD and presented to the world.

SPECIAL REFINEMENTS

In this section I explain in greater detail some of the techniques introduced in the steps above.

Mastering, like many other aspects of producing music, requires a critical ear. When you master, you can add "exciter" effects (to make some instruments or sounds more prominent in the mix), EQ, dynamics, special stereo imaging, "tube warmth," and other final touches. Good mastering plug-ins such as the iZotope Ozone 3 are quite flexible and rich in features, and, for those of us who need a little help, include lots of presets such as Country, CD Master, Symphony, Piano Rock, and so on.

Try PSP's Vintage Warmer plug-in to polish your final mix (or improve individual tracks). It helps almost every kind of music (start by using their Mastering First Aid preset for your final mix). And if you think that MIDI and computer music in general is a bit too harsh-sounding, too "digital," this unit does a great job at softening the sound, without damaging the high frequencies or otherwise subtracting from your masterpiece. Give it a look and listen at www.pspaudioware.com. The Vintage Warmer does some great things to somehow tie the sounds together, and yet give them distinct "air" so you hear each one. There appears to be some compression and limiting going on, and what other sonic fiddling they do, I don't know, but it sounds great. A similar unit I like to plug in for harder, tougher music (think "Shock the Monkey" or Mr. Reznor's pieces), is iZotope Ozone. Among other expert effects and cool presets, it can really spread the stereo field, has a fine, easy-to-use "paragraphic" equalizer, and an exciter that punches things up pretty well. And it divides the effects into various bands so you can, for example, add excitement to only a specific frequency range.

So, if you want to add some punch and polish to your version of "Bennie and the Jets," click Ozone 3's Piano Rock preset. Wow, it does add qualities that you might spend many weeks trying to achieve by fiddling with effects yourself. If Piano Rock isn't extreme enough, there are other presets to audition, such as the tasteful, though sonically violent, Punch the Monkey (though I personally feel this one should only be used with extreme caution). Download the Ozone 3 demo from www.izotope .com/products/audio/ozone/download.asp.

Adjusting the Faders

Does something that's supposed to be in the background (like bass or string pad) stand out and call too much attention to itself? Lower its volume until it blends in. Is the melody too far back, too much blended in with the support instruments? Increase its volume. The faders are there for you to manipulate until you get the right mix. If you also want to adjust the faders *while* the song plays, to bring something forward, for example, just for a few measures, record those adjustments and when you do the final playback (the mixdown to stereo), the faders will automatically move just as you recorded them. This process is called automation. Automation can be applied to volume, pan, FX parameters, mutes, and send levels.

Panning

Also while you're listening to the song prior to the mixdown, see about adjusting the pan sliders to move instruments across the stereo field between the left and right speakers. It's not common in most musical styles to bounce the instruments around while the song plays, but you do want to move them where you want them prior to the mixdown. If everything is spot in the center, what's the point of having stereo at all?

Adding Reverb

While you're at it, you can also manipulate a couple of effects to each track. Reverb is by far the most common effect—it's added to pretty much everything except sometimes percussive sounds that you want to be really bright and punchy, and of course purist electronica (computer bleeping music)

Savvy Tip

One exception to the "don't bounce" rule is the toms. It's quite common to pan the toms dynamically to give them some life during a few measures of a song. Listen to a song with toms in it and see how they move across from one speaker to the other.

Fig. 10-6: In PowerTracks you click the down arrow under the track number to drop open a window where you can adjust the track's reverb, chorus, and velocity.

where the totally dry sound is part of the appeal (these timbres don't exist in the real world, so what would the sound waves bounce off? The motherboard?).

For most instruments, though, you do want to add some reverb. It's not necessary, however, to add individual reverb to each instrument separately; you can apply reverb to the entire mix later, as if all the musicians shared the same space and, thus, were affected by the same walls. (See figure 10-6.)

Adding Chorus for Sheen

Guitars, voice, and strings particularly can benefit from some chorus—an artificial doubling or multiplying of the instrument, as if it weren't solo but rather part of a chorus of voices all trying (but failing) to hit exactly the same notes at the same time, and possibly other elements. Chorus can add sheen, body, and realism to some instruments. Try listening to a solo voice or guitar part, then add more and more chorus to hear the effect. Technically, the chorus effect is produced in a variety of ways, depending on who designed the effect. Often the signal is merely delayed very briefly then recombined with the original with random subtle variations.

Adding Other Effects

You can spend all year fiddling with the sounds of the instruments prior to mixdown. If you want to make the sound more "there" in the room with the listener, try giving it added punch by boosting

Savvy Tip

Reverb has the effect of filling out a mix, filling in the stereo field. Today's tastes generally require that snare drums, lead guitars, and lead and backing vocals get lots of reverb. Other instruments, such as bass guitar or kick drums, take up plenty of space themselves, without the help of added reverb.

the midrange or upper midrange using an equalizer. For a weird effect, add some preverb (play the track backward, recording the reverb that comes off the reversed notes, then add just the preverb to the normal track). There's pre-echo and hundreds more effects you can work with. But that's for you to explore. We need to get this song mixed down and off to the FedEx guy.

CREATING THE MIXDOWN

Good, you like what you hear. Time to send it down to a stereo track. At this point, the MIDI notation tracks (which themselves have no audio, but merely describe how a synthesizer should make music) turn into an actual audio recording. The new track will be a stereo audio track that you can save as a .wav file and burn to CD. In some sequencer software, such as Sonar 4, you must insert a new audio track at this point. In PowerTracks, you click on an empty track's circular MIDI icon next to its track number, and choose Track Type Stereo Audio from the popout menu. The MIDI symbol changes to a double wave symbol, as shown in figure 10-7.

Click the new audio track's number to select the track (it turns red). Click the << green double left arrow to move to the start of the song. One odd "feature" of the current version of PowerTracks is that it records a metronome sound along with the music. You really don't want this in the final mix, so first choose Options, Metronome, then deselect the checkbox labeled Recording. Now click the red circle to begin recording. When the song is finished, click the black square icon to stop recording.

You now see a dual waveform in your audio track, instead of the piano-roll lines in the MIDI tracks.

ADDING FINAL EFFECTS

There are many ways to add final effects, but for a beginner the easiest way to add effects to the entire final mix is to simply create a new stereo audio track, then solo and apply effects to the audio

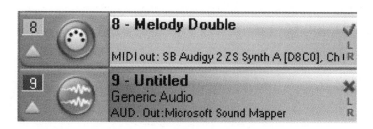

Fig. 10-7: In PowerTracks, a stereo audio track is symbolized by a double waveform.

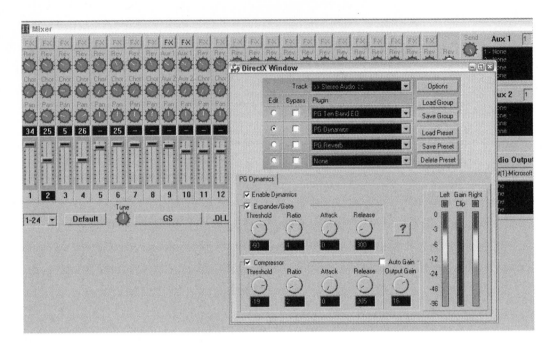

Fig. 10-8: A typical mixer gives you considerable control over various elements of the mixdown.

mixdown track you created in step 6. Click the previous track and the new track until the previous track is yellow and the new, to-be-recorded track is red. Open the Mixer window as shown in figure 10-8.

Click the FX button for the *already recorded* audio track. You want to apply some effects to that track while it plays back. You can load various effects into the dialog box shown in figure 10-8. For much contemporary music, the PG Dynamics group of effects is quite useful. Expander, compressor, and gate all in one, you can hear how these effects often improve the sound of a song. When you try it, don't be alarmed when the Clip and left and right meters (on the right side of the dynamics dialog box) go wildly into the red. This is *supposed to happen*—these aren't meters showing distortion or clipping caused by too high a recording level. So just relax and let them do their thing.

ADDING COMPRESSION

Compression is very commonly added while recording an instrument to reduce the dynamic range (the range between the softest and loudest sounds the instrument makes during recording). Additional compression is often added to the entire final mix of all the tracks. This effect is partly the result of the needs of FM radio: if you want a song to make it on the radio, you've got to ensure that the softest parts are louder (to overcome the noise of driving), and that the loudest parts don't jump out at the listener. Some people regret the pervasion of compression in today's popular music, claiming that it's often tastelessly applied and destroys the subtle dynamics in some singers' voices and some lovely songs.

EQ and reverb effects are also sometimes applied during the very last stages of the mixdown, as you can see in figure 10-8.

Doing the Master

To record the final mix (the master): Be sure that the new, empty audio track is *red* by clicking its track number in the Tracks view window. Now all you do is click the << button to ensure you're at the start of the song, then click the red circle button to start recording. Click the black square button when the song is finished. To save the resulting .wav file, be sure that the newly recorded, final mixdown audio track is *red* by clicking its track number in the Tracks view window. Then from the menus choose File, Wave Files, Export to wave file. You're done.

Of course the process outlined above is one of many dozens of ways to move from musical idea to finished song. You can record audio tracks directly from a living guitarist or songbird, right into the mix of existing MIDI tracks. Or you can edit the MIDI notes or audio waveforms using tools built into many sequencers, or via specialized editing software applications. If you're not a great instrumentalist—for example, Agnest Goeztz, whatever else she is, cannot play the oboe—try "step recording" where the time between each note recorded into the sequence track is up to you. Take all the time you need to enter each note—and if you make mistakes, you can change the notes later via an editor. There are many paths from initial inspiration to final master recording.

More about Effects

Effects are usually added to individual *audio* tracks (you normally wouldn't want to flange an entire song, but perhaps one of the tracks might benefit from it). However, when you do the final mixdown, you might want to add certain effects at that time (such as limiting, compressing, reverb, EQ). Purists will add many effects just to individual tracks, such as a particular compression that brings out the best in an acoustic guitar. This level of detail can take lots of time, though such effort can produce the best results. Also note that many effects cannot be applied to MIDI tracks, only to audio

Most quality sequencers allow you to save favorite effects settings as individual presets (your preferred settings for, say, drum EQ for reggae songs), but also groups of presets (your favorite reverb, EQ, compressor/limiter settings for, say, Very Heavy Metal).

(.wav) tracks. The few effects that can be used on MIDI tracks include arpeggiation, echo, delay, event filtering (such as adding legato to certain notes on the basis of various rules you can specify), quantization, randomization, and transposition. The majority of what most people think of as effects, however, are applied only after you've recorded, or mixed down, your MIDI track to an audio track: wah-wah, chorus, ring modulation, gates, modulation, phase shifting, and dozens more.

SEQUENCERS: GOOD, BETTER, AND BEST

What sequencer (or perhaps more accurately, *digital music production suite*) should you look at? Several sequencer packages specialize in loop-based music, the topic of the next chapter. But for straight-ahead, traditional sequencing, I'll recommend several very popular packages, starting with PowerTracks, the subject of much of this chapter.

Good: PowerTracks from PG Music

PowerTracks is a good sequencer, and for some people it's going to be all they need. Sure, it doesn't have some of the features of the more advanced sequencers (such as surround sound effects, reverb, and mixing, a new and sophisticated technology just now emerging as a result of the popularity of home theaters and their 5.1 up to the latest 8.1 speaker systems).

Perhaps you won't even miss video tracking features or advanced time scaling and dithering. Maybe the quality of the results you get isn't going to win you a Grammy next year for technical achievement. But for what you pay, I know of nothing that comes close to the rich set of features and relatively painless learning curve offered by PowerTracks. If you're a savvy music consumer interested in making the transition to music *maker*, PowerTracks is a wonderful place to start.

Better: Sonar 4 and Cubasis

When you're ready to kick in some more bucks to move on up to more control over your music, and some sonic improvements as well, I recommend you audition Sonar 4 from Cakewalk. And you might also want to check out Cubasis.

Sonar

Long a leader in computer sequencers, Cakewalk changed the name of its sequencer a while back to Sonar, but the company still does the great job of writing usable, quality software it has been doing for years. Sonar 4 actually represents a leap forward from what was already a high-quality product in earlier versions.

The Sonar line now includes several versions at different price points: Sonar 4 Producer is $600 (all prices street). The Sonar Studio version is $300, the Home Studio 2004 is $250, and Home Studio Version 2 is $100. And of course there's always eBay. Check out the features list for each version to see which suits your needs. If you're not doing video-related production and also don't expect to be getting into surround sound anytime soon, maybe you don't need the features offered in the Producer version. In any case, you can upgrade if necessary; the company has a pretty fair upgrade policy. And before you plunk down the greenbacks (or should we now say the pastelbacks?), get a demo at www.cakewalk.com/download/default.asp and give it a try.

As you can see in figure 10-9, Sonar 4 has many windows and doors behind which you'll find many useful, quality music tools. This software is very well thought out by seasoned pros who've been mak-

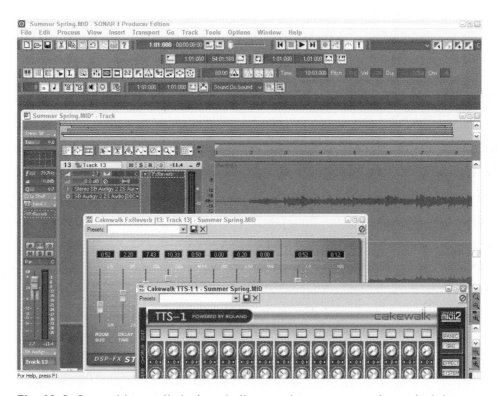

Fig. 10-9: Sonar 4 is a well-designed all-around sequencer package. And the built-in Cakewalk samples are great.

ing great sequencers and related multitrack software for many years. You get the benefit of their commitment to quality, a focus on ergonomics, and a list of features generous by any standards. The more you use Sonar, the more you appreciate the care that went into ensuring that you get great results efficiently. The product is simply a pleasure to use and learn—it's that great and rare combination of effective user-interface design with tremendous power under the hood. You know the type of software I'm talking about: things usually work the way you'd guess they will, so you learn quickly and get better results sooner than you would expect.

Cubasis: A Contender

Different folks have different ways of working with music, so no one sequencer package fits all. Good as it is, Sonar might not be what feels most comfortable to some. If you want to try another highly regarded package, consider products from Steinberg. Their Cubasis VST 4.0 is a beginner-level bundle that ships with, for example, the Garritan Personal Orchestra (described in chapter 9), or you can purchase Cubase SE (Start Edition), which is geared to those on a budget or just starting out ($95 street price). The higher level versions are SL at $250 and SX at around $600. The version I discuss here is Cubasis VST, shown in figure 10-10 (for coverage of the SX version, see chapter 13).

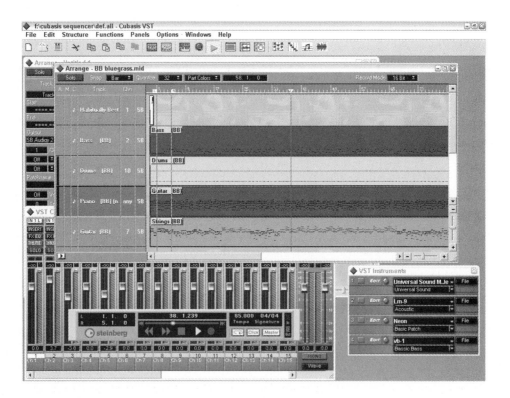

Fig. 10-10: An entry-level Cubasis VST 4.0 sequencer from Steinberg.

I found Cubasis intuitive enough, but that said, some of its ways of operating seemed to me awkward, if easily discoverable. And even given that it's an entry level sequencer, I missed some of the features found on other sequencers. Nonetheless, Steinberg has a strong reputation in the industry, and there are many who swear by the Cubase line of applications. Perhaps it would be more fair if you were to evaluate the SE version (which shares the user interface offered by the higher-end SL and SX software) rather than the Cubasis software. The Cubase SE version includes 48 audio tracks, unlimited MIDI tracks, five insert effects, and eight send effects per channel, handles up to 16 VST instruments, and ships with a collection of audio and MIDI effect plug-ins.

Best: Pro Tools

Let's say that you won the lottery, just got a fat contract, have platinum ears, or otherwise will settle for nothing but the best regardless of price. You need to look into Digidesign's Pro Tools. And at prices starting at $450 for the Mbox system and going up to around $14,000 (for the cards alone; add the 192 analog to digital I/Os for $4,000 each) for the HD 3 Accel system, many Pro Tools systems are the gold standard in every sense of the word.

Pro Tools systems combine hardware and software, thereby taking some of the burden of music production off the computer's memory and processing power and spreading it around to ancillary machinery. On the low end is the portable starter Mbox system, which includes the Pro Tools LE software, lots of plug-ins and utilities, and a hardware USB-powered Mbox with two analog inputs and outputs, 24-bit stereo digital I/O and signal path, two analog TRS inserts, a headphone jack with volume control, and zero-latency monitoring.

Pro Tools is pretty much what its name implies: the system you find in most high-end audio studios.

And, as you doubtless would expect, Pro Tools is beyond capable. It can easily accommodate enormous numbers of tracks, high throughput, multiple simultaneous behaviors, sophisticated audio-video interfaces, smooth integration with the hardware (after all, the hardware comes from Pro Tools too), and loads of top-of-the-line bundled applications and plug-ins from some of the most famous names in the business, including Live Lite 4 Digidesign Edition, Propellerhead Software Reason Adapted, and IK Multimedia's SampleTank SE, AmpliTube LE, and T-RackS EQ.

THE MISSING MACROS

Aside from some Kurzweil hardware sequencers built into keyboards, I don't know of any sequencer application popular today that includes macro facilities. And I think macros would be a significant addition. When you manipulate music in a multitrack application suite, you find yourself frequently

> **Savvy Tip**
>
> You can find general-purpose macro recording utilities that are effective with any program, including sequencers. They sit in the background, and when you invoke them (via a special hot-key combination, for example), all the recorded keystrokes and mouse moves are repeated. This can be a great time saver. I recommend Macro Express; it's full-featured and effective. Give it a try by downloading the 30-day free trial version at www.macros.com.

doing the same series of tasks, in the same order, over and over. For example, I often randomize aspects of the drum track to eliminate the machinelike perfection of the rhythm. In either of my favorite sequencer applications, this requires a number of steps involving menus, adjusting virtual "knobs," specifying percentages and "humanization" settings, and so on. If only I could have a macro record all these steps, then forever after I could just click the macro's icon on a macro toolbar and instantly all these time-consuming steps would be carried out. The best word processors and other office applications have had macros for fifteen years, and the best photo/graphics applications followed soon thereafter with their own macro facilities. Such a capability is long overdue in music-management software. Yes, I've submitted the suggestions; now I'm patiently waiting for the inevitable to, at long last, appear in somebody's feature set.

While it doesn't have an actual macro *recording* feature (though it did in earlier versions), Cakewalk's Sonar sequencer does have a built-in language that you can program to make the software do your bidding. The language is called CAL, and you can find more information about it at www.svpworld.com/cakewalk_cal/cal_part_2.asp.

> **Savvy Tip**
>
> Some of what macros accomplish can be achieved by creating templates or stationary pads, features available in many sequencers. For example, to create a stationary pad in Pro Tools, you create a new session and then save and close. On the Mac, click on the session icon and enter apple+I. This will get information about the session. There is a box to click labeled stationary pad. With this labeled, every time thereafter that you open the stationary pad, a dialog appears asking if you want to edit the stationary pad or create a new session. Similarly, in Reason you can go to the Preferences menu and create a custom session file that opens each time you run Reason. These solutions aren't as automatic as macros, nor as flexible (you can't do everything with templates), but they're a help and save you some repetitive steps.

On to Samplers

Some multitrack sequencers focus on the kind of music that uses relatively small samples, repeating them over and over, layering additional samples, adding effects, while underneath everything is a powerful beat, also known as *the groove*.

In other words, these sequencers specialize in loop-based music, improvisation, and real-time performance. Such applications often overlap the features of the sequencers covered in this chapter, but their ultimate purpose in life differs somewhat. In the next chapter we'll look at the major aspects of samplers, sample collections, and the software that, collectively, contributes to the skill known as *remixing*. It's not at all the same thing as simply creating a new, alternative mixdown.

11 REMIXING AND SAMPLING

- ◆ What is loop-based music?
- ◆ Understanding remixing
- ◆ Ableton Live 4
- ◆ Real-time sequencing
- ◆ Sony's Acid
- ◆ Steinberg's HALion SE
- ◆ All about samples

- ◆ Proteus X
- ◆ Spectrasonics's Stylus RMX
- ◆ IK's Sonik Synth and SampleTank
- ◆ Opus 1 Orchestra
- ◆ GigaStudio
- ◆ Native Instruments
- ◆ Freebies and bargains

This chapter covers two somewhat loosely related topics. Both subjects are called *sampling*, but they're really quite distinct in the way they're used. First I'll look at the latest craze—playing samples live, in real time, as if the sampler were itself an instrument. Then I'll cover the more traditional meaning of sampling—employing quality digital recordings of real (or synth) instruments using a sequencer to play the sounds.

Mixman, GrooveLab, Reason, Live, Fruity Loops, Magix, Mixmaster, Kinetic, EZ-Mixer, Stylus, Groovemaker, Acid—there's lots of good performance-oriented or loop-based software ranging from free up to hundreds of dollars. What these programs have in common is that they focus on samples and short MIDI tracks—relatively brief snippets of music that are combined, layered, overlaid on existing music, or repeated (looped) with variations, or sometimes without variations. A sample designed for loop-based music making is usually one measure long, and sounds reasonably rhythmic when repeated over and over.

My beloved older readers might say, "Why do that?" The answer can be found in two popular segments of today's music, notably hip-hop and electronica (or *house* or, if really effective and combined with lots of caffeine or other stimulants, *trance*). Some younger readers won't call these applications sample banks or sequencers at all, but rather groove boxes or something unprintable.

Compared to, say Beatles-era sound, which reached perhaps its peak of orchestral luxury in the sonic waterfalls in *Pet Sounds* and other 1960s classics, much of today's most popular music strips away melody, a good amount of harmony, even singing. Hip-hop is often spoken rather than sung, and electronica is primarily instrumental. In much of this music, the terms *monotonous* and *repetitious* are compliments.

What you're left with is the essence of music, the drumbeat, a caveman beating out a rhythm with a bone on a stone. Of course some will complain that this is the essence of music in the same way that a TV dinner is the essence of food. But tastes differ. And simple, highly rhythmic music might be just the thing at 3 a.m. when everyone's eight miles high.

The *groove* is the percussive beat, the rhythm of the piece upon which everything else is layered, *if* anything else is layered. Plus some modern twists like a synth bass line, changing tempo without affecting pitch (and vice versa), adding scratches (the only known remaining use for old-style vinyl-record players), spacey drums, warping, violent cross-fading, reversed audio, time stretching, special effects, samples from other people's music (with their permission, of course), beat bending, bizarre (relatively) synth sounds, tempo syncing, chaos funk, and, in general, beats aplenty.

REMIXING

If you're deeply into *real-time* music making, you'll want software designed for live *performance*. The sequencers explored in chapter 10 are designed primarily for traditional production environments where you, and, say, your close friend Sting, sit around and try different mixes to see which you like best. All in the leisure of a private studio on, say, Montserrat, with a jigger of rum.

But what about those who want a tool that specializes in mixing *while* people sway, or dance, to the evolving mix? What about a performance sequencer? Or if not used in actual live performance, at least one that's optimized for *remixing*?

Mixing Isn't Remixing

Don't confuse mixing with remixing. When you create a remix you're not just trying different fader settings or effects sends. You take a song, perhaps from a CD, break it down into little pieces (called samples or loops), and reorganize those pieces—under the supervision of a very dominant drumbeat. Or you might blend bits from more than one song and superimpose some original melodic ideas of your own creation. The most common tactic is to fracture, then recombine, a song into something creative, original, and, above all, with a good groove.

Ableton Live

One such program that offers features optimized for real-time looping is the appropriately named Live 4 from Ableton ($400 street price). You can play this thing like an instrument, though there's certainly quite a bit you can do to prepare *before* the rave party as well. And professional DJs do prepare in the same sense that professional musicians always prepare: they practice, get their musical elements together, and otherwise visualize, or to an extent even rehearse, the performance well in advance of the big night. Of course, with this kind of music, improvisation plays a big part too.

Distinguishing between traditional sequencers and these new performance sequencers isn't entirely black and white. It's a matter of emphasis—what features the manufacturer brings to the front. Think of Live and similar sequencers as improvisation—or even performance—tools. They're designed to be primarily *instruments*, with recording and editing features available. Think of sequencers like Sonar (see chapter 10) as a recording and editing tool, with some improvisation tools included. It's not that Sonar doesn't have facilities for manipulating loops; it does. It just doesn't expect you to want to manage them in all kinds of ways rapidly in front of hundreds of bouncing ravers. Unlike Live, Sonar isn't a *sequencer as instrument*.

One of the first things you notice when you dive into Live's (and other performance-oriented sequencers') approach to music is that a song is divided into little pieces, what Live calls *scenes*, such as small loops, brief verses and choruses, and perhaps an intro and end. Each of these relatively small sections can be played *in any order* by simply clicking a scene-launch button in Live. In other words, you can rearrange the song on the fly by launching scenes in varying orders: the music isn't set down in the sequencer as a linear piece of music but instead is broken into pieces (the scenes) that can be dynamically selected by the composer during the performance of the piece.

A song is even more deeply deconstructed because each scene can be subdivided into *clips*—the individual instruments playing in the scene. You can combine clips—for example, replacing the bass line in chorus 1 with the bass line in verse 2. This kind of micro mixing works best, of course, if the song is in one key and one tempo, but much of this kind of music is. And these types of sequencers

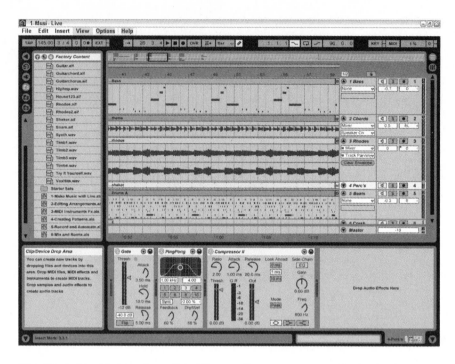

Fig. 11-1: Here's Live in action. Three effects are operating in the bottom middle panel, while the tracks above them are recorded.

have special features that you can use to dynamically alter the tempo, key, timbre, and other elements of the music on the fly as well. In other words, you get little pieces of music (brief audio samples or MIDI loops) and combine and modify them dynamically as they play. Replacing loops or recording new samples is quite simple. And replacement loops (or even *audio* samples) can be forced to automatically sync with the song's tempo, so you don't get any bizarre, unpleasant counter-rhythms that would doubtless bring most of the dancers out of their trance. Always remember the DJ's motto: The beat must go on. Figure 11-1 shows you how Live looks to a frenetic DJ.

One of the coolest features in Live is called *warping*. You set one or more warp markers within a sample (see figure 11-2), and then you can stretch or shrink the sample, distorting its feel and rhythm in (sometimes) interesting ways. This technique can also fix the tempo if Live incorrectly estimated it when automatically syncing it with the rest of the music.

If you're just starting out as a dance music producer or DJ, Live includes a good set of easy-to-follow tutorials to get you up to speed quickly.

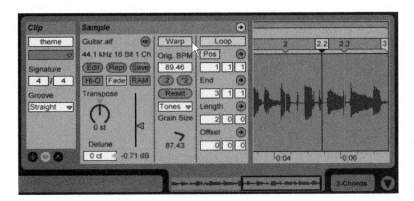

Fig. 11-2: Live's warp feature allows you to drag a portion of a sample, modifying its tempo without changing the pitch.

Sony's Acid

Another contender for best of category in the performance sequencer division is Sony's Acid Pro 5 ($250 street) (see figure 11-3). This full-featured multitrack package beefs up VST support, now including VST effects, Rewire, and multiple VST instruments. You get three impressive Native Instruments' software synths (described later in this chapter) and the Sound Forge Audio Studio application (a home version of the professional application, and a solid audio recording and editing utility).

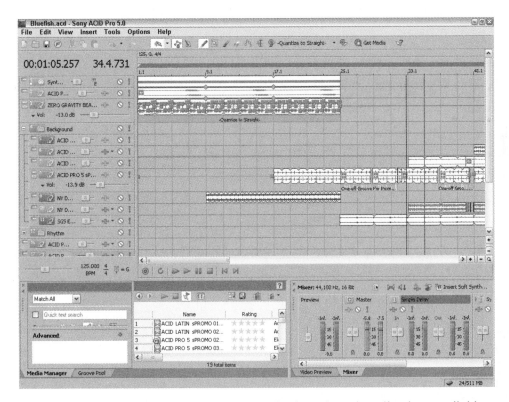

Fig. 11-3: Acid Pro 5 is one of the most popular loop-based applications available.

Acid is a powerful application, though some find it more difficult to use than Live. Both packages have their strong points, and both are great at letting you quickly build a new song idea and save a demo to let others see what they think. Which software package is most comfortable for you is always a matter of personal feel and your approach to music, so compare these and other performance sequencer packages before making your final choice.

One interesting feature in Acid is the way you can import an audio track from a CD, then let Acid's Beatmapper Wizard help you sync it to the tempo of your current project. Try this:

1. Choose File, Extract Audio from CD. The Beatmapper utility appears, offering to figure out how to sync this audio track to the tempo of your existing project.

2. Click Next.

   You're shown a waveform of the stereo audio, and a line through the waves displays what the Wizard thinks is the start, the downbeat, from which it should calculate and adjust the tempo. You're asked to listen and verify that this is, in fact, the downbeat.

3. Click Next, and now you're asked to verify the length of a measure in this song. You can click the play button in the Wizard to hear the audio played along with Acid's metronome. You further inspect measures to ensure accuracy, as shown in figure 11-4.

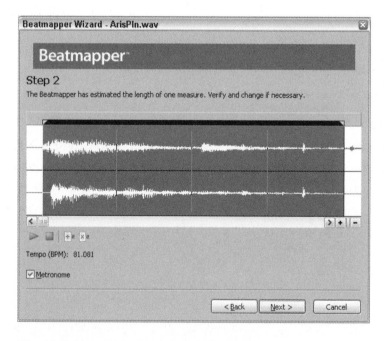

Fig. 11-4: Use Acid's Beatmapper Wizard to match imported audio CD tracks to your current project's beat.

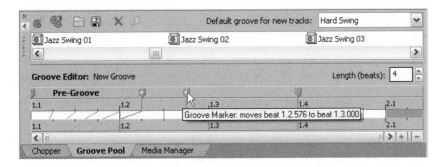

Fig. 11-5: Slide these groove markers to modify the beat in Acid Pro 5.

Not to be left behind technologically, Acid includes a somewhat different version of Ableton's sample warping capability. Called *groove mapping* in Acid Pro 5, it allows you to manipulate a track's rhythm, as shown in figure 11-5. You can create groove maps from whatever sample you wish, then modify them or map them to other samples. Adjusting the rhythms of samples becomes quite easy.

UNDERSTANDING TRADITIONAL SAMPLING

If you plan to work with musical samples in a sequencer, and in the comfort of a quiet room or studio (as opposed to playing loops live in a steaming nightclub), you're talking about *traditional* samples.

You can buy samples by the cartload (see the section below on Stylus) or you can record your own, copy them from existing CDs (that aren't, of course, copyrighted), get them on DVDs included in computer music magazines, or download them from the Internet.

A sample is usually a relatively short audio recording, but sometimes a brief loop is also referred to as a sample. Strictly speaking, a loop is usually longer than a sample. A loop is composed of, say, several samples arranged together in a particular rhythm that can be repeated and remain in sync. But the terms are frequently interchanged. A sample can be a recording of something quite brief, like a snapped finger, or something many measures long, like a sustained grand piano note. But we'll start with a sample collection that's nontraditional and is designed to work with an application like Live.

Spectrasonics's Stylus RMX

The Stylus RMX ($275 street price) is a plug-in, sample-based virtual instrument that you use with a host sequencer. Stylus is focused on electronica / urban remix music, to be played in real time, so it's not really a traditional sampler. It includes loads of sounds (7.4 GB: over 10,000 single hits, 2,500 groove elements, and 3,000 kit modules) with charming names like The Outlaw, Atomic Zoo, and

Fig. 11-6: Select your samples and grooves in Stylus's browser view.

Raw Meat. You know, the kind of music Doris Day used to sing. They're pumped through the Spectrasonics Advanced Groove Engine (SAGE) for instant response. Stylus is optimized for performances but can obviously be used for remixing in the studio as well. Among its other talents, Stylus might just be the ultimate drum machine. (See figure 11-6.)

Stylus always plays its samples in sync with the sequencer it's running under, so no matter what the original tempo of a groove or sample, it's adjusted to match the master tempo of the entire project. (The samples' tempos are conveniently listed in the librarian, but you can ignore them if you wish; they'll be made to sync if you load them in.)

If you're familiar with synthesizers, you'll be comfortable with Stylus's many ways to shape sound: reverse, filters, LFOs (low-frequency oscillators), pitch and envelope adjustments, and so on. Figure 11-7 shows the sound editor window.

One interesting feature in the Stylus suite is the Chaos Designer, a full-featured tool that introduces the unexpected and adds variety to your music. Move the sliders, shown in figure 11-8, to add surprise random behaviors to Pattern, Repeat, Reverse, Timing, Pitch, and Dynamics.

Fig. 11-7: Here's where you can modify a sample in many ways, the same way you'd noodle around with a hardware synth to get the perfect sound.

Fig. 11-8: Use the Chaos Designer to shake up your music and eliminate predictability. You can add subtle or wildly obvious randomization.

IK's Sonik Synth and SampleTank

If you're after a highly regarded sound library for *all* genres of music—not just electronica and hip-hop—you should consider IK Multimedia's Sonik Synth 2 ($320) (see figure 11-9). You get over 5,000 sounds and 7 GB of samples, editing via 50 synth/sampler controls, 32 digital signal processing effects (you can use dozens of effects per sound, five insert effects per layer), and more. It's also compatible with its sister software, SampleTank. Sonik Synth's samples are for the most part excellent; many are gorgeous. And you'll find all kinds of variety ranging from keyboards, choirs, acoustic basses, guitars, and drums to electronic instruments and famous legacy synth sounds, along with special sounds like wineglass hits and float-on-a-cloud pads. You can freely blend sounds, such as layering acoustic sounds with synthetic sounds—up to 16 layers. Sonik Synth 2 also boasts one of the largest collections of vintage synth samples ever released in one product. Not only does it include relatively typical minimoog, Jupiter, Prophet 5, OB-Xa, and ARP 2600 samples, but it also contains rarer instruments such as the EMS SYNTHI Modular, Moog Modular, Gleeman Pentatonic, Prophet T-8, and many others. Think of it as a massive synth museum—and a real resource when you're looking for that perfect sound.

SampleTank

SampleTank is similar to Sonik Synth but contains its own unique set of samples. However, the sample system is completely compatible with Sonik Synth—so if you own SampleTank 2 and want to

Fig. 11-9: Sonik Synth, IK's mega sample library and synth workstation

Fig. 11-10: SampleTank2 includes lots of sounds, permits real-time editing, and shines as a virtual playback synth.

expand your sound library by 7 gigs, buying Sonik Synth can be an affordable upgrade path. SampleTank comes in two main VST versions: XL ($380 street) and L ($230) (see figure 11-10). The high-end XL version has 96 acoustic, 82 orchestral, and 77 electronic sounds—a total of over 4.5 GB. The L version is 2 GB. SampleTank is a sample playback unit (and you can import samples into it), not a recorder or editor, but it does its job very effectively. If you're a fan of its sounds, and the idea of synths in general, you're likely to enjoy many included sounds as is, or you can fiddle with them using the plentiful built-in effects and lots of synth-style adjustments such as envelopes, filters, and LFOs.

Sonik and SampleTank are similar, though there are differences. Both are sample synths with a wide variety of sounds, and both support AU, DX, and RTAS technologies to VST.

A key distinction—beyond their unique sample libraries—can be found in the fact that SampleTank includes the feature set found in Sonik Synth but also adds the ability to import new samples (.wav, .aiff, akai, and samplecell). You can build your own sample library and also purchase "expansion tanks." Then you can use some of the great tools included in SampleTank (such as Stretch, which ensures that timbre is preserved as a sample is played at different pitches) to modify any of the samples. And of course the interfaces look different. Perhaps you'll want to try them out. The SampleTank demo can be downloaded from www.sampletank.com/Main.html?STFreeDwn.

You can find a demo of the new Sonik Synth 2 at www.soniksynth.com. Or give it a try at your local music store—if it *carries* computer music. Some salespeople look astonished if you ask about computer software, in much the same way that one imagines their ancestors dropped their jaws if you inquired about a *piano* during the last years of the reign of the harpsichord. As we all know, *real* musicians never get anywhere near a computer.

My only real problem with IK's SampleTank and Sonic is the size of some of the typefaces; I find them a bit small, so I wish there were an option to adjust them. Of course, you get more information on screen thanks to this size, but permitting user preference (as they do with the color of the instrument) would have been nice.

Garritan Personal Orchestra

If you want a reasonably priced collection of classical orchestra instrument samples, most of which sound to my (and others') ears exactly like the real thing, take a look at the Garritan Personal Orchestra (described in detail in chapter 9). These instruments aren't, of course, limited to classical music. Lots of rock songs—and even some techno—can benefit from the occasional timpani hit or authentic-sounding Steinway grand. You wouldn't catch Elton John playing a synth piano, would you?

Opus 1 Orchestra

If money is no object and you want 25 GB of truly astonishing classical orchestral instrument samples, take a look at the Vienna Horizon Series Opus 1 Orchestra (or one of their other sample packages). You'll want to have GigaStudio to use these samples with your sequencer, and you'll also need a sound card compatible with GigaStudio. But, what the heck? Hey, big spender: If you've got the cash to buy into these fabulous samples, what's a few more hundred to support them by upgrading your system?

Listen to some of the demos at www.etcetera.co.uk/products/VIE010.shtml.

GigaStudio

While we're on the subject of GigaStudio, you should also give it an audition. It uses a wonderful technology that streams samples off your hard drive rather than storing them in RAM. Given that most computers have .5 GB of RAM storage but over 100 GB of hard drive space, you can see the advantage. The samples streaming off a hard drive can be much, much larger (meaning the notes can be held longer without having to loop them, compress them, or otherwise compromise their quality). GigaStudio can be purchased in three versions—Solo, Ensemble, and Orchestra—ranging from $160 up to $499. I've had some problems with stability in XP Service Pack 2, but this issue will probably have been solved by the time this book is published. Why bother? Because the sampler is feature rich and includes some of the best samples you'll ever hear.

Native Instruments

No overview of today's best samplers and sample collections would be complete without mentioning the products from well-regarded Native Instruments. You can buy their entire bundle, the whole enchilada—NI Komplete 2—for $1300. If that cash isn't lying around the house, you can try some of their individual samplers and specialized products. You can get their excellent version of the classic Yamaha Dx7—the first synth I ever owned, and a great technology in its day. NI's version allows you to reproduce the entire range of cool (some would say *cold and hard*) sounds from the original hardware synth. But it improves on the original by permitting you to add considerable warmth, and fatness, to the Dx7's notoriously clinical timbres. Thus, NI's FM7 Software Synth allows you the option of accurate Dx7 sounds or improvements on them. Expect to pay around $200, but, as always, check eBay.

Additional high-quality products bundled in Komplete 2 that you might like to check out individually are NI's Absynth 3, B4 Organ, Battery 2, Intakt, Kompakt, Kontakt 2, Main Pro 53, Reaktor 4, Spektral Delay, and Vokator. Komplete 2 includes every NI sample engine, synth, and effect—plus four sample libraries—in a single bundle. But you can certainly purchase them individually if you prefer.

FREEBIES AND BARGAINS FOR THOSE WITH MORE TALENT THAN MONEY

You can find many of my favorite commercial products described throughout this book. I've generally found them reliable, feature rich, stable, well supported, often generously endowed with bundled additional software: overall a solid investment. The cost of computer music has nose-dived and the features have greatly increased in the past few years. But, good as professional commercial products can be—all those sample libraries, sequencers, effects and other plug-ins, soft synths, ROMplers, utilities, and so on—you can also find some great inexpensive shareware (sometimes freeware) sounds and music software.

Your starting place to look for downloadable plug-in goodies is www.kvraudio.com. It's a great site where you'll find lots of audio plug-ins employing the common technologies: VST (platform independent), DirectX (Windows only), Audio Units (Mac), and LADSPA (Linux). The site also includes informative user ratings of both commercial and shareware plug-ins, and other features of interest to musicians. It's your one-stop shop for low-cost, high-quality computer audio applications, utilities, reviews, and other digital music stuff.

Plug-ins need something to be plugged *into*—a host application, usually a sequencer. If you're on a budget and have lots more talent than money (a temporary embarrassment soon to be rectified, I'm sure), you can't get a better value for your money than the free recorder/editor application Audacity. Take a look at screen shots and the features list at audacity.sourceforge.net/about.

Sometimes you get lots more than you pay for. And even if you can afford other audio software, you might still find yourself using Audacity. I do. For one thing, it offers excellent *what you hear* recording capabilities. With this feature, anything that comes out your computer's speakers can be captured as a high-fidelity stereo digital .wav recording.

12 CHOOSING THE RIGHT GEAR FOR A LISTENING ROOM

◆ Sound conditioning

◆ Rearranging a room

◆ Testing the sound

◆ Soundproofing

◆ Buying speakers

◆ Putting the *psycho* in psychoacoustic

◆ Finding a receiver

◆ Final tweaking

If you've decided to dedicate one of the rooms in your house as a listening room/home theater, you'll find some useful ideas in this chapter. We'll explore your options when buying hardware gear such as speakers, a receiver, cables, a power supply, curtains, and so on.

Curtains? Yes, often the biggest single improvement you can make to the quality of your home audio is to make changes to the listening room itself. Fabric absorbs sound (refuses to reflect it like a bare wall does). So when you're designing a music-listening room, or a home recording studio, you need to take into account various acoustic issues, not least of which is reflectivity. Adding some heavy curtains, moving the couch around, and other maneuvers can have a major impact on the sonic quality of your room.

Let's see what can be accomplished with a music-listening room, or—what amounts to essentially the same thing—a home theater. Just as a DVD player can both show movies and play CDs, so too can a home theater do double duty for both video and audio. And the surround sound typical of DVDs is now becoming increasingly available on CD—currently via SACD and DVD-Audio. Although this technology has yet to catch on with the music-listening public at large, it seems inevitable that relatively soon *stereo* recordings will fade into history.

DEAD ROOMS, LIVE ROOMS

Fortunately, one of the primary ways to optimize a home theater—particularly if you're using a large-screen TV—is to cut down on the ambient light. You want to hang curtains that block the light coming in through the windows. This avoids screen reflections and improves perceived brightness and color in the video image.

This is fortunate because curtains usually also improve the sonic qualities of a room. Most music is carefully recorded and mixed to optimize what the artist and producer think sounds best. This includes adding a "room"—an interior space represented by the reverb that's added to the sound, and also perhaps some natural ambience created by where the sounds were recorded.

In other words, its creators cannot predict how you will hear their music (in what kind of room, in a car, via headphones). So the idea is that your listening room should be neutral, relatively dead, sound-absorbing. Your room shouldn't add its own, unpredictable reverb, echo, standing waves (sound at frequencies relative to room dimensions that gets amplified via reflections), or other sonic side effects, such as a loose window rattling in tempo with the bass drum.

That's the theory, anyway. But in practice you could spend thousands of hours and dollars creating the perfect audio-video room, only to put on a CD that has been optimized for noisy-ambience car radio playback or mixed to sound good in the perfectly neutral sonic environment of headphones.

SOUND CONDITIONING: ADD CARPET AND CURTAINS

Even though you won't achieve audio reproductive perfection this side of heaven, there *are* some practical things you can do. Perhaps the most efficient and impressive improvements can come from adding fabric. Get a thick carpet to avoid standing waves between the ceiling and floor.

And avoid having two reflective (painted, wallpapered, mirrored, or otherwise flat and hard) walls

Pick out a test CD. Choose a CD with music you love and know well. Decide how you like that CD to sound, then aim for that when testing audio equipment you're thinking of buying, or when experimenting with your listening room's acoustics.

facing each other. For example, if you have a wood-paneled wall, hang heavy curtains on the opposite wall. If one wall of the room is hard plaster and the opposite wall is all picture windows, cover the picture windows with a nice thick curtain (gauze won't do it) that you close when listening to music or watching video. And if possible, listen in a room large enough that you don't sit too near the walls.

Plump, plush furniture (think a high-back, soft fabric chair, or, alternatively, a mesh-backed chair like the Aeron chair by Herman Miller) also helps prevent unwanted sonic reflections, as do irregularly shaped rooms—though the latter can have unpredictable effects. Your final decision on how to arrange furniture in the room depends on several factors, but for audio purposes, try moving things around, installing a heavy carpet pad, and the other tactics mentioned above, all the while listening to a favorite test CD before and after. Listen to see if you can hear the various instruments separately; if they sound as they do when live; if there are any strange booming or thumping sounds in the bass range; if you hear too much echo—or any other annoying sounds the room is adding to the CD. Then listen to that same CD via headphones, to see what happens when you *subtract* the room. (But be aware that headphones tend to distort the stereo spread, sending it right *through* your head.)

Another important aspect of room tuning is bass buildup. Bass buildup can occur in corners and severely distort the perception of the audio. Many companies build foam bass traps that sit in the corners. You can purchase these at any pro audio dealer or build your own. Corner sonic traps can also help a square or rectangular room tighten up because they will stop much of the sound that finds its way to the corners. Visualize how frequencies bounce and build energy in the corners, for the same reason that you cup your hands around your mouth to yell at someone to amplify your voice. This corner effect is well worth avoiding.

Arranging a Room to Suit Your Own Tastes

Approaches to listening room design come and go. There have been fads and fashions over the years. And experts still disagree about what makes the perfect listening environment. Many suggest that no walls be parallel at all, and that the ceiling be high and sloping. They go on to recommend wall treatments of rubber, egg-crate-shaped foam, cork, and so on. Some insist that all walls be covered

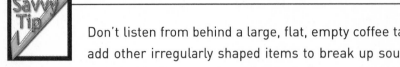

Don't listen from behind a large, flat, empty coffee table. Pile some magazines on it and add other irregularly shaped items to break up sound waves bouncing off it. You know how some people hold silver panels under their face at the beach to concentrate the sunlight and speed their tanning? Coffee tables can have that effect on sound, and to most people it's not a desirable effect.

in the *same* nonreflective, but simultaneously nonabsorbing, material (what qualifies for *this* specification? Kryptonite?).

Some experts say you must *not* use a high-back chair—you'll block the rear reflections necessary to give music "dimension" and "air." Some people won't like the effect produced by any of the suggestions in this chapter. There are no hard-and-fast rules—if you love highly reflective (very "live") environments, perhaps none of the deadening suggestions above will suit your taste.

Work toward what pleases your ears. Experiment. After all, hi-fi was originally an effort to reproduce, with great fidelity, the original sound of, say, the violin, or a singer's voice. But as technology has improved, the goal of many sound engineers, singers, composers, and others has moved away from fidelity as a primary objective. They're not so much aiming for what sounds real as for what sounds *good.* And increasingly those goals diverge.

Singers are frequently doubled, to give their voices an unnatural, but pleasing, depth. Similarly, chorus and other effects are added all over the place to please the ear. I won't claim that we've reached the point of decadence represented by epicurean feasts. Those nasty Romans sometimes told the cook that everything should be transformed from its natural origins (fish disguised to look like birds, birds disguised to look like mice, and so on). But the traditional goals of high fidelity no longer apply in many recording studios.

Testing a Music Room

Back to your music room tests. Listen particularly for clarity in the bass range. Do you hear a variety of bass timbres, or is one frequency dominating (as is typical of cheap car stereos—thump, thump, the same thump over and over)?

Now turn up the volume, let the subwoofer go boom. Does anything start buzzing in the room—pictures rattling, pipes vibrating in the walls? You can, if necessary, dampen vibrating objects with felt, wire, screws, whatever. If this extra sound doesn't bother you, well, dude, then don't bother fixing it.

Maybe the buzz only happens when cannons are fired in a DVD movie anyway. Or maybe you like the effect and think it's—as your grandfather would have said—*snazzy*.

Stand in the middle of the room and clap once loudly. Can you hear an echo? If so, you may want to further dampen the room. Too much damping, though, can result in an unpleasantly bland sound. So maybe you'll want to remove a couple of those new velvet curtains you just installed. It finally comes down to your taste, and of course the kind of music you like.

Listen to see if the vocals and high-frequency instruments are clean sounding or whether there's too much reverb from the room (compare the room sound to the same music played through headphones).

SOUNDPROOFING: DOUBLE THE WALLS

Soundproofing is quite a different matter from the acoustic conditioning described above. The idea with soundproofing is to create a sound barrier between your room and the outside world. This has two benefits. First, it can prevent your beautiful listening experiences from being interrupted by a passing boombox car or Baby Becky's crying. Second, it keeps *your* sound in the room so you can crank it up guiltlessly—without worrying about who you might be bothering elsewhere in the house, or beyond.

You don't have to pour a concrete bunker, though that *is* pretty good insurance against transmitting vibrations.

Ideally, you'd simply float your listening room, making it independent of the surrounding house via rubber hockey puck–like discs or some other isolation devices. Sound waves can be blocked, just as jetties block ocean waves.

For most people, however, true soundproofing is too much trouble and too expensive. For the best isolation, you need to build a second room inside your existing music room. This is rather more than most of us can manage, but it's pretty much what you have to do to get really good results. Build new walls inside the existing walls, but use a different frame—don't attach the inner wall to the same studs as the outer wall. Then stuff this wall sandwich you've created with insulation, such as fiberglass, along with gypsum or other sound-deadening materials. Details on constructing such a room within a room can be found in the excellent *Handbook for Sound Engineers* by Glen Ballou (Howard Sams, 1991).

If these ideas are impractical for you, there are still a couple of things most of us *can* do to improve, if not greatly eliminate, sound transmission in and out of the listening room. Loose doors and windows act as panel speakers when loud sounds—particularly at low frequencies—bounce against them. Replace single-pane with double-pane windows and install a *solid* heavy door. Use weather stripping to seal that door so it fits tight in its frame.

BUYING SPEAKERS

Huge amounts of time, research, money, and magazine space are devoted every year to speakers (they used to be called *loud*speakers, for reasons unknown). In college I bought some small Japanese bookshelf numbers that sounded great in the store in LA. Did the salesman mess with the EQ in the store's receiver to appeal to the who-needs-midrange, naive ears of a youngster? Probably. In the same way, projection TVs' "standard" video settings produce an eye-popping brightness, sharpness, cool white, and contrast that attempt to make them stand out from the crowd of other TVs in the store.

Finding Dream Speakers

Twenty years later, after I had a best seller and was flush for a time, I got my dream speakers: tall electrostatic beauties that looked like Arthur Clarke's monolith, costing around $1,500. They were great for 12 years until Tiger, my sweet husky, decided in a moment of weakness to "mark" one of them. After that, it sounded like music piped through Rice Krispies.

The quest for perfect sound never ends. I've seen pictures of big, gorgeous, solid stainless steel speakers poised like space aliens on three gleaming cone legs. A pair of these costs more than a luxury car. If I had the cash, I'd buy them for their beauty as sculptures alone.

As for studio standard: for many years the Yamaha NS-10m's were the "cheapo" reference speaker used in almost all recording studios in the world. You will know these by the ubiquitous white cones. These speakers did not sound great, and they were fatiguing on the ears, but engineers noted that when used to mix down a song, the NS-10m's produced a mix that sounded very good on "regular" home stereos. The venerable NS-10m's are no longer available new, but a pair can be found as little as $250, making them a great entry-level studio monitor speaker. It's always important to monitor your music with a view to the target audience (auto radio? home hi-fi? iPod?).

A new studio standard that can be found in a great many studios is Genelec speakers. These speakers sound great and are relatively easy on the ears. They are also powered by their own internal amplifier (so you don't have to worry about matching the amp to the speakers). However, if you are looking for a very good powered speaker and the Genelecs are out of your price range, you might want to give powered Mackie studio monitor HR824 a listen.

Watch That Loudness Button

My advice? Take your test CD to the store, ensure that no equalization (see if the "loudness" button is pressed in—it emphasizes the highs and lows at the expense of the midrange), "Mega" bass boost, or other trickery is going on while you compare speakers. And if they don't sound good in your room at home, exchange them for another set. (Can't say "pair of speakers" any more because car and home "stereo" these days usually involves at least six speakers, including the subwoofer. Actually, shouldn't say *stereo* either.)

SPEAKER PLACEMENT

Where should you put the speakers in the room? Here, too, you'll find conflicting advice. Most experts agree that putting a speaker in a corner of the room creates a megaphone effect— it amplifies the bass. Where the experts disagree is whether this is desirable or an unpleasant honking effect you should avoid. As usual, try it both ways and see what you think. My feeling is that the corner placement isn't really a good thing. Instead, use a subwoofer to supply accurate low-bass sound.

Don't let your speaker cabinets themselves vibrate—put three (not four) pieces of felt in a triangular pattern under each speaker to lift it just enough off the floor so it cannot rat-a-tat-tat, adding a surprise new instrument to the band on the CD. Or try some commercial speaker stands.

STRANGE TWEAKS

Some people fall in love with audio. And audiophiles, like all lovers, enjoy buying presents for the object of their desire. You don't even have to go to the high-end audio/video salon to find some of these expensive gifts. Stores like Circuit City offer special "tuned" cables with "no loss" gold tipples, "smoothed" power supplies, and so on. Some people swear that they can hear a difference when they use more expensive connectors, cables, and other accessories. Others, myself included, have sometimes bought them but can't claim to notice any improvement. On the theory that nothing's too good for my baby, I got some *really thick* cables with *really strong* insulation. After all, if the wire is thick, the music will flow through it faster and easier, right?

I've read about people (who claim to hear a difference) sometimes failing supervised A/B listening tests, but who am I to stop somebody from enjoying their hobby? As for psychoacoustics, I think that some of these products fall more into the *psycho* than the *acoustic* category. But I suspect that people who say they hear a difference are sincere. Even placebos have a real effect. And a really sturdy

> **Savvy Tip**
>
> Cable connectors need not be gold, but do rub them with fine sandpaper now and then to ensure that the plugs and the sockets they plug into are clean and free from oxidation. And the cable wire itself shouldn't be extremely cheap and thin. Wire the size of electric cord is usually thick enough. As a general rule, the longer the distance the cord has to go, the thicker it needs to be.

lightning, spike, and surge protector power supply is just common sense after you've invested a lot in your audio/video hardware.

Also consider purchasing an APC Battery Backup. You plug your computer and other important gear into the battery backup, and if the power goes down, the battery comes on and you can turn your computer off after you save your project, instead of allowing the lightning to turn it off without a save.

BUYING A RECEIVER

Today's receivers are all-in-one home entertainment centers: an audio/video switcher (though you cannot usually route HDTV through them—the studios are afraid to let consumers get their hands on a purely digital video stream, so that's been blocked), AM/FM tuner, amplifier, surround sound processor, and sometimes some digital effects thrown in, such as reverb and bass effects. Here's what to look for:

- Get a well-known brand, such as Sony, Onkyo, Denon, or Harmon/Kardon. Steer clear of "bargain" no-name units like *Wazoo*. If you want more rarefied audiophile gear, read reviews of NAD, Rotel, and others described in vivid, loving detail in such publications as *Stereophile Magazine*.
- Look for at least 100 watts per channel for a listening room of average size, and at least 5.1 channels.
- Check epinions.com and online reviews to see which current models most please the public and the critics.
- Get all the 5.1 surround sound technologies—Dolby (both Pro Logic II and Dolby Digital), THX, and DTS (and Neo 6)—because you don't know what software might appeal to you. If you're going high-end, you can look for 6.1 (THX Surround EX, Dolby Digital EX, DTS ES Matrix, DTS ES Discrete 6.1). Or for 7.1, THX Ultra2 Cinema mode and MusicMode.

> **Savvy Tip**
>
> The smaller your speakers, the more power they need to produce a given volume level. And if you turn up the volume on a low-power receiver, it can go into distortion and fry your speakers real fast.

If you're planning to get hardware dedicated for studio use alone (as opposed to also serving your general audio needs—games, personal hi-fi listening, movie surround sound, etc.), you might want to consider buying a professional audio amplifier rather than a receiver. Consumer receivers include a plethora of options not necessary for good studio monitoring. For the same price as a consumer receiver, you could instead buy a Hafler 1500 or 3000. These amps sound great, are silent, and are built to operate all day, every day for many years.

Also, consider auditioning some of the many low-cost powered speakers now available for studio use. These speakers take a lot of the guesswork out of matching an amp with speakers—the amp's already inside.

Tweaking

After you've assembled all the hardware and furniture and you've wired everything together, it's time for a final hour or two of tweaking before you're done. You have to ensure that the speakers are all in phase and that the sound level is balanced, and make a few other little adjustments.

In my view, the easiest way to fine-tune your system is to purchase a copy of *Video Essentials,* a DVD that allows you to optimize both your audio and your video settings. It lets you test connections, surround balance (via pink noise), center channel polarity, and so on. You can find it on Amazon.com and in some audio stores.

13 SETTING UP A PRO-QUALITY, ALL-DIGITAL STUDIO FOR UNDER $1,000

◆ Understanding what equipment you need

◆ Starting with the computer

◆ Adding software

◆ Plugging in a sound card

◆ Fixing up the studio

◆ Customizing the sound

In the past few years, professional-quality music studio equipment prices have not just dropped, they've plunged. This chapter puts it all together, showing you how to set up a home studio that rivals what pros could only wish for just a few years ago.

The big news in music these days is the democratization of studio time. It used to cost thousands an hour to get near-professional recording studio gear. Now for less than $1,000 you can own a studio that's fully capable of recording, mixing, and pressing CDs of the best possible quality—all inside your computer. Whether you're putting out a demo or selling your own CDs after local gigs, you no longer need a recording contract to record first-rate music CDs.

What accounts for this shift from the expensive, huge pro installations common a decade or so ago to today's studio-in-a-software-box? You guessed it: the personal computer. Just as everyone can now run a photo lab in their computer, so too can musicians buy software that does nearly everything recording studio hardware does, except for some analog I/O peripherals: mics, keyboards, monitor speakers, and some actual real traditional instruments and singers (if your musical inspirations require them).

WHAT YOU NEED

The following is a list of essentials for a recording studio, though if you want to pursue perfection—at least until you exhaust your budget—the sky's the limit on the price you can pay for equipment.

The Computer

For your computer, you need a relatively new machine. Digital audio recording (and mixing) is a demanding endeavor, requiring a fast CPU, lots of RAM memory, and a big, fast hard drive. Windows XP offers more software and hardware options, but many creative people like the Mac; 512 MB RAM is good, Pentium 4 running at 2.4 GHz or better is desirable, and a CD burner, and a large-capacity, fast hard drive (over 100 GB) are essential. This chapter's title assumes you *already* have this machine. If you need to buy it, cross out the "$1,000" and replace it with "$2,000."

The Software

You can get an all-in-one "studio-in-a-box"; I recommend either Cakewalk's Sonar 4 Producer ($600 street) or Cubase SX ($600)—see chapter 10. Each has excellent facilities for all the steps in digital music making: the initial recording, editing (if necessary, correcting bad notes, rhythmic oddities, whatever), and mixing. If you can get by with fewer cutting-edge features (surround reverb, video capabilities, and so on) take a look at the next-lower level of software, such as Cakewalk's Sonar SE (Studio Edition). You can still get brilliant results with software that's not the absolute high-end version. Check out the companies' Web sites for video and audio demonstrations of the software, as well as actual demos you can download and interact with yourself—for a while.

After you've got the finished audio tracks on your hard drive, software to do your final burn to a CD is relatively straightforward. I use Nero Express (www.nero.com/us/index.html).

As with most things audio, opinions about what's "best" differ. This book's tech editor—who has lots of experience in professional music studios—makes the following alternative suggestions:

Tech Editor's Note: As an avid user of Pro Tools (I own an Mbox, 002r and 32-input TDM Mix+++ system), I'd like to recommend an alternative starting place for a home studio. In my experience, an Mbox running the free, bundled Reason, Ableton Live, and SampleTank software is about the best thing running for under $450. This takes care of the audio interface (including 2 Focusrite preamps, digital in and out, etc.), the software for Pro Tools, the other free bundles, and the sound card for the computer. You get all the tools required to create incredible MIDI scores using Reason, mix in some live loops from Ableton, and record it into Pro Tools digitally. Then aspiring musicians can take what they have done and seamlessly transfer it to a professional studio, which, without a doubt, runs Pro Tools.

The Sound Card

A sound card is a peripheral that connects the outside world to your computer, providing inputs for mics or electronic instruments such as MIDI keyboards, outputs to an amplifier and speakers or headphones, and perhaps some onboard digital signal processing to take some of the load off your computer's CPU. The sound card supplied with your computer (sometimes no more than a chip on the motherboard) is generally inadequate for the demands of digital audio recording and mixing. For a good sound card, you can expect to pay a minimum of around $200, and it can go up to the thousands.

The sound card you use for any MIDI input must support ASIO (audio streaming input/output). Many of today's sound cards do include drivers for Steinberg's ASIO technology. A sound card manufacturer must write a driver (writing drivers is heavy-duty programming) that employs the ASIO technology, bypassing the built-in, multipurpose audio system in Windows and ensuring low latency. Latency is the amount of time it takes a note to sound after you've pressed a key on a MIDI keyboard (or other input device). It can be very difficult to play if there's a considerable delay—even a fraction of a second can be disconcerting. It's like you're hearing an echo, but not the original note.

M-Audio has a respected line of cards, as do Digidesign, E-Mu, MOTU, and others. If you're just getting started, you might try Creative's 2ZS Platinum Pro or 4 Pro models. The 4 Pro costs about $275 street and the 2ZS Pro is $50 less. The 4 Pro includes one additional stereo input (six inputs versus the five inputs on the 2ZS) and what's described as improved digital to audio converters and signal to noise ratio. As always, before you buy, check reviews and customer opinions.

Searching the Internet for *sound card* might be too limiting. Also look for *control interface, MIDI controller,* and *audio interface.* Or ask Sweetwater or some other musician's resource to send you a catalog (go to sweetwater.com).

New products are coming out monthly as digital audio explodes and more and more people realize they can now accomplish what they've only dreamed of: achieve professional recording results. The balance has shifted. It's no longer true that your talent is being limited by your hardware. Rather, in many cases the hardware will be limited by the talent. Not you and me, of course. Those others.

There's another step you should take to ensure you're getting the peripherals you need to interface the outside analog world (mics, monitor speakers, acoustic guitarists) with the digital studio inside your computer. Go to a large music store that sells software (i.e., is computer savvy). Describe what you plan to record. The store's computer guy can often take you right to the card or outboard interface you need. But do visit more than one store, consulting with more than one computer specialist, if possible. And some online stores like Sweetwater will give advice over the phone. You want to get objective recommendations. And *always* research purchases online by Googling for *"product-name* review" and visiting epinions.com.

Miscellaneous Instruments and Hardware

If you're going for live recording—as opposed to inside-the-machine looping, for example—you need some additional hardware to input the music into your computer's recording software. You need whatever MIDI instruments, mics, and cables are required to achieve the sound you're after. For a more complex studio setup, you may want a hardware mixer to route the various incoming signals to the computer. Alternatively, some sound cards, such as M-Audio's Delta line, can route multiple inputs to the virtual mixer in your Sonar, Cubase, or other software DAW (digital audio workstation).

FIXING UP THE ROOM

If you intend to record live performances, you should condition a room, including soundproofing to prevent birdsong, crickets, sirens, and other outside noise from wrecking a great take. Of course, some music isn't harmed by environmental sounds—there is such a thing as "live" recording. And some types of music, such as electronica, can be entirely sampler/synth based. In those cases, noise pollution is impossible because the sounds are never outside the computer in the first place. Nothing gets recorded via a mic. But you may need to isolate the musicians from outside sound (see chapter 12 for suggestions about constructing a room-within-a-room that's acoustically decoupled from the world).

As for improving your recording sessions, much depends on what kind of music you work with. One consideration is *leakage*—one instrument being picked up by another instrument's mic. For example, the tracks get a little out of your control for later mixdown purposes if the vocalist is being picked up by the drum mic.

You can record each instrument at a separate time, of course, to prevent leakage. But many bands depend on vocal and visual cues from each other, and they get a different result. They often just sound better when recorded ensemble.

Savvy Tip

One musical effect that I've always admired is a sudden solo—it can lift a song up into the air. For example, the band is playing along but during mixdown you suddenly mute everything except the drummer for a few measures. But if you've got leakage, you can't do this because the drummer's mic was picking up the vocalist, faintly but audibly. Likewise, if you decide to rerecord the original vocal track, you'll be fighting against the competing leaked vocal on the drum track. (Leakage prevention can get pretty complex, particularly if you're multimiking a drum kit.)

The least expensive solution to leakage and other unwanted sonic debris is to *totally* deaden the recording studio. I'm talking wool blankets everywhere, or some other highly absorbent material. You record the tracks as dry as possible and add reverb and other effects when you mix the music down. Of course during recording, you must either use headphones or isolate monitor speakers in a totally separate "control" room, to ensure that they're not providing massive leakage into the recording microphone.

As for the environment during mixing itself, I believe there's no substitute for headphones. No room is sonically perfect, so you just eliminate the room entirely by putting the headphone's speakers right up against your ears. Sony MDR-V6s are great headphones, though they can be difficult to find because they're often sold out.

Building a New Room

If you're handy and plan to construct your own studio or refurbish a garage, most of the music listening room suggestions in chapter 12 apply to a recording studio as well. You can find recording studio plans and instructions on the Internet at sites such as www.ibiblio.org/studioforrecording/cottage.html and www.ethanwiner.com/acoustics.html.

I agree with those who say you should make the walls irregular, if possible—no wall should face any other wall. The most attractive solution is to build more than four walls. Also, it's a good idea to slope the ceiling so it doesn't reflect sound off the floor. These tactics discourage standing waves.

Baffles and Gobos

You can spend eternity tuning a room for optimum open-mic and acoustic-instrument recording. If you want to get further into refining the sonic quality of your room and avoiding leakage, vibration, rattling, and other defects, search the Internet for *bass trap, acoustic foam, studio baffle, acoustic panel, gobo, recording air gap,* and *acoustic diffuser*. You'll find plans so you can build these items yourself if you're on a budget. But as your research will show, experts disagree about the best approaches to creating recording studios. However, there does seem to be a growing consensus that rigid fiberglass absorbs sound better than foam.

With the pointers and techniques described in this chapter and elsewhere in this book, you should be well on your way to taking your music to the limit of your ability.

GLOSSARY

aftertouch—Quality **MIDI** keyboards detect what your finger does following the attack, or initial hit on the key. Experienced musicians can use their fingers to manipulate the sound, adding realism in various ways depending on which instrument they're playing. Violinists, for instance, often move their fingers to cause the sound to waver in pitch (*see* **vibrato**), or a guitarist slightly bends the note. Aftertouch can usually be programmed to produce filtering, pitch, vibrato, loudness, or other changes.

arpeggiator—A feature used in a **sequencer** (either hardware or software) to play the individual notes of a chord separately, rather than all at once. To use the C major chord, for example, an arpeggiator plays C then E then G rather than playing CEG simultaneously. If you hold down CEG with the arpeggiator active, it can play runs all the way up and down the keyboard—à la Liberace—if you so request. Arpeggiator behaviors can be adjusted if you want an arpeggio that's a little less flamboyant.

ASIO (audio streaming input/output)—A special technology created by Steinberg to avoid the **latency** caused by slow built-in Windows audio systems.

audio card—AKA *sound card*. The computer's sound-handling peripheral, most often with features for playing and recording sounds. Most cards today include quite a few additional features, such as digital signal processing (**DSP**) and **MIDI** capability; many even feature ASIO.

audio file—A computer file such as .mp3, .wav, aiff, sdII, or .wma containing sound. It can be a short sample, a sound effect, a song, or a full orchestral movement. It can be compressed or uncompressed. (*See* **compression, MP3, WAV, WMA**.)

audiophile—Someone who loves music, especially someone who is interested in maximizing the quality of his or her equipment as a demonstration of that love.

bit depth—Along with **sample rate**, a way to specify the quality of digitally recorded music. Bit depth refers to the number of bits used to store each sample. More is better. The bit depth is 16 bits for CD, allowing a number ranging from 1 to 65537 to be stored for each sample. SACD (super audio CD) and DVD audio generally employ 24 bits, or a range of nearly 17 million. Any kind of sonic manipulation—remixing, adding **reverb**, whatever—adds errors into the digital domain. Having a greater bit depth reduces the impact of these errors when the signal is eventually restored to analog form and pumped as waves out speakers to our ears.

bouncing—Mixing several **tracks** into one, usually to free up space for additional recording. This used to be a necessity when recorders and **mixers** had fewer tracks available. It's increasingly less significant as the computer takes over recording and mixing jobs that were previously done via hardware.

bpm—Beats per minute. Most sequencers express tempo by this measurement.

burning—Recording something on a CD.

channel—Often a synonym for **track**. When used to describe MIDI, there are 16 channels for **MIDI** data. In a **mixer**, it's the strip of **faders**, effects knobs, **VU meters**, and other tools that control a single track of music.

chorus—A widely used audio effect that doubles a sound (such as a violin) but delays one of the layers slightly. Chorus fattens the sound, adding sheen. You have to hear it to understand it. Most **sequencers** have a chorus feature, so turn the knob up on a **track** and listen to what it does for a voice or musical instrument track. It sounds as if two or more instruments are playing in unison. Some synths reduce the pitch just a bit in addition to adding a delay. *Chorus* can also mean a simulation of an actual group of human singers. This effect is achieved using multiple, varying delays in addition to a bit of feedback. Yet a third meaning of *chorus* is the repeated portion of a song.

clipping—Digital distortion due to overload (usually indicated by red lights on the **VU meter**).

compression—An **effect** used during music recording that prevents the loudness of a signal from varying too much (above a specified level), which is known as dynamic range. Used particularly with popular music. Compression reduces dynamic range. Often employed with a **gate** and a **limiter**. Compression also means reducing the size of a computer data file while, presumably, retaining the quality of the original recorded audio. **MP3** is a popular compression technology.

DAW—Digital audio workstation. A complete recording-studio-in-a-box. Everything you need to record, mix, and otherwise manipulate music. It can be a hardware unit, but it is increasingly an all-purpose software application, such as Sonar 4. Sometimes the term **sequencer** is used interchangeably.

DirectX—Microsoft's programming interface that allows more efficient access to low-level multimedia functions (such as audio).

driver—Sophisticated software that communicates between hardware like a sound card and an application such as a **sequencer**.

dry—See **wet**.

DSP—Digital signal processing. Digital manipulation of audio, such as adding **reverb**.

DXI—A software synth **plug-in** technology developed by Cakewalk. It employs **DirectX**.

effects—Changing the sound of audio. **Chorus** and **reverb** are the two most common effects. Special effects (or **FX**) is a term also applied to somewhat bizarre sounds, such as sirens or thunder.

enhancer—A self-congratulatory name for an **effect** that, something like **chorus**, is intended to improve the psychoacoustic quality of a singer or other instrument. It often involves **real-time** manipulation of phase, equalization (**EQ**), faux harmonics, and other factors. Try it, you might like it. You might even think it *enhances* the sound and justifies its name. Unlike chorus, though, it's not a standard effect on digital software **sequencer** mixers. *See* **exciter**.

EQ—Equalization. Like a tone control, but more sophisticated. Instead of merely adding or subtracting bass, treble, and so on, an EQ effect unit is divided into perhaps ten or more "bands" so you can selectively affect the sound. A *parametric* equalizer allows you control over the amount of boost or cut, the bandwidth, and the center frequency.

exciter—A specialized kind of **enhancer** that primarily spends its time adding harmonics that were not present in the original sound. Again, try it to see if you like the effect. Be warned: it may cause people like Kramer to go bonkers, so don't overdo it.

expander—Used to widen the dynamic range by decreasing the softer sounds and increasing the louder sounds. It's the opposite of a compressor (see **compression**).

fader—Usually a slider (a potentiometer) that adjusts the volume of a **track** on a **mixer**. Controls the "gain," the volume.

flanger—An **effect**, a delay that employs feedback. Sounds like rotating speakers or something.

FX—effects, like **reverb** and **flanging**.

gate—an **effect** that silences (or reduces gain of) a channel if the signal strength falls below a specified level. Often used along with **compression** to improve the dynamics. A gate can remove noise.

HDCD—High-definition compatible digital, a CD that sounds better than ordinary CDs. Traditional CDs employ 16-bit word length; HDCD boosts this to 20 bits, resulting, proponents claim, in better spatial placement, increased dynamic range, and more accurate, natural-sounding instruments. Alternative "superior sound" CD formats include SACD (super audio CD) and DVD-audio. SACD is capable of carrying multichannel audio, and uses a technology named Direct Stream Digital.

humanizing—See **quantizing.**

latency—A delay between when you press a key on a synth keyboard (or other input device) and when the sound actually plays. Latency is highly annoying and can make it very difficult to stay on the beat.

leakage—The sound of one instrument getting into the mic (and thus on the recorded **track**) of another instrument. This is highly undesirable because if, for example, the drum leaks into the singer's track, you cannot then easily edit the singer's track. And what if you then decide to rerecord the drum part? You'll have a problem with that leaked drum sound that's still on the vocal track.

limiter—Enforces a specific loudness level and refuses to permit the sound to go above it. *See* **compression**. A limiter is the opposite of a **gate**.

loudness control—A button or knob that boosts bass and treble frequencies, leaving the midrange as is. At low volume, the ear is less sensitive to low and high frequencies, so this simple **EQ** adjustment helps compensate. Audio dealers are notorious for using this button, though, to make certain (high-profit) speakers sound "better" to novice listeners. You know, "deeper bass" and "more highs."

mash-up—The latest craze: blending two musical genres, via sampling, into a single song.

mastering—Mixing down multiple **tracks** into the final stereo pair of tracks (or, lately, into the final five, six, or seven surround tracks), just prior to **burning** to a CD.

melisma—That really annoying vocal trick of singing every note *except* the correct note, until finally settling on the right note. Some singers warble around a single syllable, floridly adding their own "music" to the original melody. It's the vocal equivalent of break dancing—startling at first, but quickly exasperating because it's tedious, limited, and ultimately unaesthetic.

MIDI—Musical instrument digital interface. The standard language or protocol by which electronic instruments communicate with each other and computers. Notes, instruments, **pitch bend**, and many other aspects of music are each given specific codes. Music can be very efficiently stored in MIDI and reproduced precisely through a MIDI-capable device such as a sound card (*see* **audio card**) or **synthesizer**.

MIDI cable—A wire with MIDI plugs on each end. Used to connect MIDI devices.

MIDI event—A single MIDI command, such as "note off."

mixer—A software or hardware device that allows you to merge two or more **tracks**. You can also at the same time manipulate **effects**, volume, and panning.

modulation—Usually controlled by a wheel on a **MIDI** keyboard. The modulation wheel can be set to control various **effects** such as adding **tremolo**, **vibrato**, **pressure**, and so on.

MP3—A music storage scheme that compresses audio files (*see* **compression**). It's the current, and likely future, standard for efficiently storing music in computers or portable music devices.

multitrack—a **mixer**/recorder (software or hardware) system that permits individual instruments to be recorded separately, each on its own **track**. This way, you can later add additional tracks, or edit individual tracks without affecting all the other tracks.

normalizing—Recording at the highest volume setting that doesn't result in distortion—getting the lines bouncing high in the **VU meter**, but not so high that they get into the red zone. You usually play the music through your **mixer** before recording, just to check the input levels.

orchestration—Choosing which instruments to use for a musical piece.

panning—Moving sound between the stereo pair of speakers or, with **surround sound**, placing a sound somewhere within the sound field. A **mixer** has a pan knob for each **track**.

patch—a **MIDI** instrument, such as a harmonica patch.

pitch bend—Adjustments made to the pitch, usually via a wheel controller found on **synthesizers** or other electronic keyboards. **MIDI** can record pitch bends, along with most other elements of a musical performance.

pitch to MIDI converter—Software that listens to an audio (.wav) (see **WAV**) file or live instrument input and attempts to translate the sounds into **MIDI**. This almost always fails to achieve very reliable results. If you have a monophonic signal, with very discrete, pure tones, it's still going to make mistakes. So, as with handwriting recognition, the technology has a way to go before you can MIDIfy any old No Doubt record. It will be nice, though, if the technology ever succeeds.

plug-ins—Utilities or other add-on programming that runs "within" another application, such as a **sequencer**. In other words, a plug-in adds a feature to an application, such as **surround sound** mixing, **samples**, a vocoder effect, and many other features. Saying that it runs "within" it means that you activate the plug-in from a menu in the host application, and the plug-in directly interacts with the host, seeming to simply be another feature of the host.

podcast—From *iPod*; podcasting does for Internet music what TiVo does for television: it allows time-shifting. A listener can receive a podcast for later listening during his or her commute. Listeners don't have to do anything special to get a podcast "show" downloaded; it can be automatically sent. Subscribe, and your machine ensures you get any new shows on the "stations" you're interested in. If you want this, download software from http://ipodder.org/directory/4/ipodderSoftware.

portamento—Enforced gradual changes from one pitch to another, even though you're pressing keys (for example, on a **MIDI** keyboard) abruptly.

presence—The sense that music is being played right there in the same room with you. An **audiophile** term of endearment. Highly subjective.

pressure—Similar to **aftertouch**; sophisticated keyboards detect how your finger presses the key *while* the note is being played, and produce **vibrato**, **tremolo**, added brightness, or volume.

quantizing—Enforcing the rhythm of a piece by having the computer move notes so they fall on the beat (your choice of resolution—to the nearest eighth note, for instance). Quantizing can also be used to add "swing" or randomly move notes *off* the beat by a specified amount. Minor rhythmic errors are typical of human players, so when you use quantizing to induce errors into a computer-perfect (and robotic-sounding) rhythm, you're said to be *humanizing* the music. Some sequencers and **DAWs** use the quantize feature to both quantize and humanize, but others have a separate humanize feature.

RAM—Random access memory. The erasable memory in a computer; 512 MB of RAM is considered a minimum for today's computer audio software, and 1024 MB is preferable.

real time—As time normally passes. In other words, you can record a piece by playing notes on a **MIDI** keyboard step-by-step without regard to how much time passes between notes (this is called step-recording or **step-time**, and the computer ensures that the notes will play back in tempo later). Or you can record in real time—at the speed at which the listener will later hear the music. In other words, time as most of us experience it in reality.

reverb—Ambient reflected sound. Whatever you hear after the musician stops playing. In other words, echoes and reflected sounds that depend on the environment. (Reverb is actually composed of many closely spaced echoes.) A school gym—with no curtains, soft chairs, or other sound-dampening fabrics is highly reflective and is said to be "live" because it provides lots of reverb. Reverb is one reason people like to sing in the shower. Reverb is almost always added—along with **chorus** effects—to contemporary popular recordings.

ripping—Recording audio from a CD to a computer's hard drive.

SACD—Super audio CD. *See* **HDCD**.

sample—An audio recording of an instrument (or brief piece of music, called a *loop*). Samples, if well-recorded, can sound just like the original instrument, because they *are* the instrument's sound. Synths generate sound using artificial means; samplers play an actual audio recording of a real instrument.

sampler—A specialized device or application designed to record, play back, organize, and manipulate **samples**.

sample rate—The number of times per second that a digital **sample** is taken during recording. More is better. (AKA *sampling frequency*.) Think, for example, of the number of individual photos (samples) that must be taken per second for a movie to reproduce a smooth motion picture. It turns out that the number is around 24, but 30 is better. Too few samples and you get flicker, from which comes our word for movies: *the flicks*. When sampling audio, the more often you sample, the more realistic and accurate the result. Of course, there's a point beyond which additional samples per second add nothing to the quality as perceived by human ears (which is, after all, the point, right?).

sequence—A sequence is essentially a **track**—usually a recording of a single instrument. For example, if you have a keyboard **synthesizer**, you can select the trumpet voice and play taps while the software **sequencer** listens and records your performance onto that instrument's own track or sequence.

sequencer—Originally an application that held musical information as **MIDI** tracks (or sequences) and that you used to record or edit those **tracks**. Today, however, most sequencers do pretty much everything that a full recording studio does—**mixing**, adding **effects**, holding audio tracks (in addition to MIDI tracks), and so on. *See* **DAW**.

sound card—*See* **audio card**.

step-time—*See* **real time**.

streaming—Audio that comes in from the Internet in **real time** (like radio). The alternative is to download an **audio file** and listen to it after the download is complete.

surround sound—Using more than two speakers, so sounds can come from the sides and behind the listener. Common in today's movie and home theaters, increasingly common in music listening rooms.

sustain—To hold a note. The right pedal under a piano is called the sustain pedal, and it keeps the dampers raised off the strings, so notes slowly decay naturally.

synthesizer—Software or hardware, this device generates musical instrument sounds electronically. And they usually *sound* electronic, though sometimes synth sounds can be lifelike. Synth sounds are heavily used in today's techno and electronica music genres.

tempo—The speed at which music is played, expressed as beats per minute (bpm).

timbre—The characteristic, unique sound of each instrument. For example, if both a trombone and a trumpet play middle C, you can still tell them apart. What distinguishes these sounds is the instruments' timbre.

track—Generally a single instrument is recorded to a single **track** (in **multitrack** recording). However, an **effect**,

MIDI controller information, lyrics, or other data can be on their own tracks as well. A **sequencer** typically has twenty-four or forty-eight tracks available, but some of today's **DAWs** can have hundreds of tracks.

transpose—To change the key. Often you need to lower **MIDI** tracks by an octave (for reasons nobody can explain). You transpose it by -12.

tremolo—Repeated, relatively rapid changes in the loudness of a sound. (*See* **vibrato**.)

VBR—Variable bit rate. Allows an increase in the **sampling rate** as needed to ensure high quality when using **compression** to make a digital recording. If you specify a minimum bit rate of 128kbs, for example, and also use the "best quality" setting for your recording, you're likely to get a result that varies—as required—between 128kbps and 160kbps or so. The alternative is a constant bit rate, and lesser quality.

velocity—How fast, thus how hard, you hit a key on a **MIDI** keyboard, resulting in a louder, or sometimes brighter, note.

vibrato—When a singer or instrument changes the pitch up and down around the actual note. Vibrato is a regular, rapid, up-and-down repetition of two notes—commonly heard, for example, when a violinist or guitarist rapidly quivers her fingers on the strings. Vibrato is often a very pleasant sound, in contrast to the annoying vocal stunt called **melisma**. (*Also see* **tremolo**).

VST—Virtual studio technology. A system of instrument sound generation created by Steinberg that operates in **real time**. VST instruments are added to **sequencers** or other music applications the same way as **plug-ins**. They're modular—you can add or remove them to manage the **timbres** and instruments available for your composition. *See* **DXI**.

VU meter—Volume unit meter. Normally two bars that respond to changes in volume in each stereo channel. You watch that the input doesn't drive these bars into the red, indicating that the sound is too loud, resulting in distortion.

WAV—A computer file format (.wav) for recording, storing, and playing back audio (as opposed to compressed or **MIDI**) music. *See* **compression**.

wet—With added **reverb** (or other **effects**). As opposed to *dry* or "dead" sound that has nothing added to the instrument's original sound.

WMA—An **audio file** format. Microsoft's answer to **MP3**. WMA manages to compress twice as effectively as MP3.

workstation—A hardware or software central controller, having a variety of functions and features available at once in one unit. *See* **DAW**.

INDEX

ABOUT THE AUTHOR

Richard Mansfield loves music. Computer music in all its aspects is one of his hobbies. He has been using hardwware and software synthesizers, multitrack recorders, mixers, and all kinds of music software for 20 years, since the early days of Cakewalk Version 1. He believes that his primary strength as a writer is explaining technology to the average person, empowering people to get involved and benefit from the amazing things computers can do. From 1981 through 1987, he was editor of *Compute!* magazine. He has written hundreds of magazine articles and two columns. From 1987 to 1991 he was editorial director and partner in Signal Research and began writing books full-time in 1991. He has written 38 computer books since 1982.

He lives in High Point, North Carolina. His books have been translated into 12 languages and have sold more than 500,000 copies worldwide.